Wireless Networking Technology

From Principles to Successful Implementation

Steve Rackley

AMSTERDAM • BOSTON • HEIDELBERG • LONDON
NEW YORK • OXFORD • PARIS • SAN DIEGO
SAN FRANCISCO • SINGAPORE • SYDNEY • TOKYO
Newnes is an imprint of Elsevier

Newnes is an imprint of Elsevier
Linacre House, Jordan Hill, Oxford OX2 8DP
30 Corporate Drive, Suite 400, Burlington MA 01803

First published 2007

British Library Cataloguing in Publication Data
A catalogue record for this book is available from the British Library

Library of Congress Cataloguing in Publication Data
A catalogue record for this book is available from the Library of Congress

ISBN 13: 978-0-7506-6788-3
ISBN 10: 0-7506-6788-5

For information on all Newnes publications
visit our website at www.books.elsevier.com

Printed and bound in Great Britain

07 08 09 10 11   10 9 8 7 6 5 4 3 2 1

**Working together to grow
libraries in developing countries**

www.elsevier.com | www.bookaid.org | www.sabre.org

ELSEVIER   BOOK AID
International   Sabre Foundation

Wireless Networking Technology

From Principles to Successful Implementation

Contents

Introducing Wireless Networking

Development of Wireless Networking

Although the origins of radio frequency based wireless networking can be traced back to the University of Hawaii's ALOHANET research project in the 1970s, the key events that led to wireless networking becoming one of the fastest growing technologies of the early 21st century have been the ratification of the IEEE 802.11 standard in 1997, and the subsequent development of interoperability certification by the Wi-Fi Alliance (formerly WECA).

From the 1970s through the early 1990s, the growing demand for wireless connectivity could only be met by a narrow range of expensive hardware, based on proprietary technologies, which offered no interoperability of equipment from different manufacturers, no security mechanisms and poor performance compared to the then standard 10 Mbps wired Ethernet.

The 802.11 standard stands as a major milestone in the development of wireless networking, and the starting point for a strong and recognisable brand — Wi-Fi. This provides a focus for the work of equipment developers and service providers and is as much a contributor to the growth of wireless networking as the power of the underlying technologies.

While the various Wi-Fi variants that have emerged from the original 802.11 standard have grabbed most of the headlines in the last decade, other wireless networking technologies have followed a similar timeline, with the first IrDA specification being published in 1994, the same year

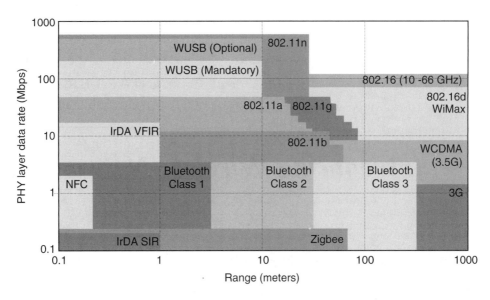

Figure 1-1: Wireless Networking Landscape (rate vs. range)

that Ericsson started research on connectivity between mobile phones and accessories that led to the adoption of Bluetooth by the IEEE 802.15.1 Working Group in 1999.

During this period of rapid development, the variety of wireless networking technologies has expanded to fill the full range of requirements for data rate (both high and low), operating range (long and short) and power consumption (low and very low), as shown in Figure 1-1.

The Diversity of Wireless Networking Technologies

Wireless networks now operate over four orders of magnitude in data rate (from ZigBee at 20 kbps to wireless USB at over 500 Mbps), and six orders of magnitude in range (from NFC at 5 cm to WiMAX, and also Wi-Fi, at over 50 km).

To deliver this breadth of capabilities, the many companies, research institutions and individual engineers who have contributed to these developments have called into service a remarkable range of technologies; from Frequency Hopping Spread Spectrum, the inspired World War II invention of a film actress and a screen composer that is the basis of the Bluetooth radio, to Low Density Parity Check Codes, a breakthrough in high efficiency data transmission that lay gathering dust for forty years after its development in 1963 and has proved to be an enabling technology in the most recent advances towards gigabit wireless networks.

Technologies that started from humble origins, such as OFDM — used in the 1980s for digital broadcasting, have been stretched to new limits and combined with other concepts, so that Ultra Wideband (UWB) radio now uses multi-band OFDM over 7 GHz of radio spectrum with a transmitted power below the FCC noise limit, while OFDM combined with Multi-Carrier Code Division Multiple Access is another gigabit wireless network enabler.

Techniques to satisfy the every growing demand for higher data rates have gone beyond the relatively simple approaches of shortening the time to transmit each bit, using both the phase and amplitude of the carrier to convey data or just using more radio bandwidth, as in UWB radio, and arrived at the remarkable concept of spatial diversity — of using the same space several times over for concurrent transmissions over multiple paths — as applied in MIMO radio.

This fascinating breadth and variety of technologies is the first motivation behind this book, which aims to give the reader an insight into these technologies of sufficient depth to gain an understanding of the fundamentals and appreciate the diversity, while avoiding getting down to the level of detail that would be required by a system developer.

As well as seeking to appeal to the reader who wants to gain this technical insight, the book also aims to use this understanding of the principles of wireless networking technologies as a foundation on which, a discussion of the practical aspects of wireless network implementation can be grounded.

Organisation of the Book

This book is arranged in seven parts, with Parts I and II providing an introduction to wireless networking and to wireless communication that lays the foundation for the more detailed, technical and practical discussion of the local, personal and metropolitan areas scales of wireless networking in Parts III to V.

Part I — Wireless Network Architecture — introduces the logical and physical architecture of wireless networks. The 7 layers of the OSI network model provide the framework for describing the protocols and technologies that constitute the logical architecture, while wireless network topologies and hardware devices are the focus of the discussion of the physical architecture.

Some of the key characteristics of wired networking technologies are also briefly described in the two chapters of Part I, in order to provide a background to the specific challenges addressed by wireless technologies.

In Part II — Wireless Communication — the basics of wireless communication are described; spread spectrum, signal coding and modulation, multiplexing and media access methods and RF signal propagation including the important topic of the link budget. Several new or emerging radio communication technologies such as ultra wideband, MIMO radio and Near Field Communications are introduced. Part II closes with a similar overview of aspects of infrared communications.

Part III — Wireless LAN Implementation — focuses on what is perhaps the most important operating scale for wireless networks — the local area network. Building on the introductory description of Part I, local area wireless networking technologies are reviewed in more detail — including the full alphabet of 802.11 standards and enhancements. The practical aspects of wireless LAN implementation are then described, from the identification of user requirements through planning, pilot testing, installation, configuration and support.

A chapter is devoted to the important topic of wireless LAN security, covering both the standards enhancements and practical security measures, and Part III closes with a chapter on wireless LAN troubleshooting.

Part IV — Wireless PAN Implementation — takes a similar detailed look at wireless networking technologies on the personal area scale, including Bluetooth, wireless USB, ZigBee, IrDA and Near Field Communications. The practical aspects of wireless PAN implementation and security are covered in the final chapter of Part IV.

Part V — Wireless MAN Implementation — looks at how the metropolitan area networking challenges of scalability, flexibility and quality of service have been addressed by wireless MAN standards, particularly WiMax. Non-IEEE MAN standards are briefly described, as well as metropolitan area mesh networks.

The practical aspects of wireless MAN implementation are discussed, including technical planning, business planning and issues that need to be addressed in the start-up and operating phases of a wireless MAN.

Part VI — The Future of Wireless Networking Technology — looks at four emerging technologies — namely wireless mesh routing, network independent handover, gigabit wireless LANs and cognitive radio — that, taken together, look set to fulfil the promise of ubiquitous wireless accessibility and finally lay to rest the recurring technical challenges of bandwidth, media access, QoS and mobility.

Finally Part VII — Wireless Networking Information Resources — provides a quick reference guide to some of the key online information sites and resources relating to wireless networking, a comprehensive listing of acronyms and a glossary covering the key technical terms used throughout the book.

PART **I**

WIRELESS NETWORK ARCHITECTURE

Introduction

In the next two chapters, the logical and physical architecture of wireless networks will be introduced. The logical architecture is introduced in terms of the 7 layers of the OSI network model and the protocols that operate within this structure, with an emphasis on the Network and Data Link aspects that are most relevant to wireless networking — IP addressing, routing, link control and media access.

Physical layer technologies are introduced, as a precursor to the more detailed descriptions later in the book, and the physical architecture of wireless networks is described, focussing on wireless network topologies and hardware devices.

At each stage, some of the key characteristics of wired networking technologies are also briefly described, as a preliminary to the introduction of wireless networking technologies, in order to provide a background to the specific challenges addressed by wireless technologies, such as media access control.

After this introduction, Part II will describe the basic concepts and technologies of wireless communication — both radio frequency and infrared.

CHAPTER 2

Wireless Network Logical Architecture

The logical architecture of a network refers to the structure of standards and protocols that enable connections to be established between physical devices, or nodes, and which control the routing and flow of data between these nodes.

Since logical connections operate over physical links, the logical and physical architectures rely on each other, but the two also have a high degree of independence, as the physical configuration of a network can be changed without changing its logical architecture, and the same physical network can in many cases support different sets of standards and protocols.

The logical architecture of wireless networks will be described in this chapter with reference to the OSI model.

The OSI Network Model

The Open Systems Interconnect (OSI) model was developed by the International Standards Organisation (ISO) to provide a guideline for the development of standards for interconnecting computing devices. The OSI model is a framework for developing these standards rather than a standard itself — the task of networking is too complex to be handled by a single standard.

The OSI model breaks down device to device connection, or more correctly application to application connection, into seven so-called "layers" of logically related tasks (see Table 2-1). An example will show

Table 2-1: The Seven Layers of the OSI Model

Layer	*Description*	*Standards and Protocols*
7 — Application layer	Standards to define the provision of services to applications — such as checking resource availability, authenticating users, etc.	HTTP, FTP, SNMP, POP3, SMTP
6 — Presentation layer	Standards to control the translation of incoming and outgoing data from one presentation format to another.	SSL
5 — Session layer	Standards to manage the communication between the presentation layers of the sending and receiving computers. This communication is achieved by establishing, managing and terminating "sessions".	ASAP, SMB
4 — Transport layer	Standards to ensure reliable completion of data transfers, covering error recovery, data flow control, etc. Makes sure all data packets have arrived.	TCP, UDP
3 — Network layer	Standards to define the management of network connections — routing, relaying and terminating connections between nodes in the network.	IPv4, IPv6, ARP
2 — Data link layer	Standards to specify the way in which devices access and share the transmission medium (known as Media Access Control or MAC) and to ensure reliability of the physical connection (known as Logical Link Control or LLC).	ARP Ethernet (IEEE 802.3), Wi-Fi (IEEE 802.11), Bluetooth (802.15.1)
1 — Physical layer	Standards to control transmission of the data stream over a particular medium, at the level of coding and modulation methods, voltages, signal durations and frequencies.	Ethernet, Wi-Fi, Bluetooth, WiMAX

how these layers combine to achieve a task such as sending and receiving an e-mail between two computers on separate local area networks (LANs) that are connected via the Internet.

The process starts with the sender typing a message into a PC e-mail application (Figure 2-1). When the user selects "Send", the operating system combines the message with a set of Application layer (Layer 7) instructions that will eventually be read and actioned by the corresponding operating system and application on the receiving computer.

The message plus Layer 7 instructions is then passed to the part of sender's operating system that deals with Layer 6 presentation tasks. These include the translation of data between application layer formats as well as some types of security such as Secure Socket Layer (SSL) encryption. This process continues down through the successive software layers, with the message gathering additional instructions or control elements at each level.

By Layer 3 — the Network layer — the message will be broken down into a sequence of data packets, each carrying a source and destination

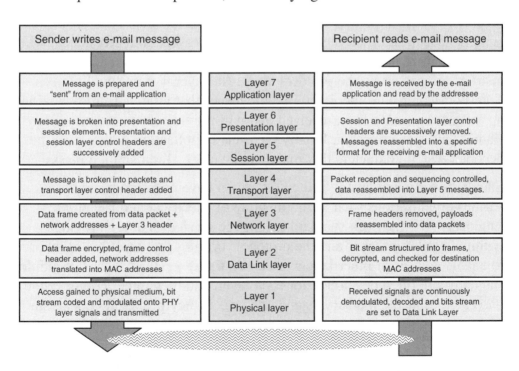

Figure 2-1: The OSI Model in Practice — an E-mail Example

IP address. At the Data Link layer the IP address is "resolved" to determine the physical address of the first device that the sending computer needs to transmit frames to — the so-called MAC or media access control address. In this example, this device may be a network switch that the sending computer is connected to or the default gateway to the Internet from the sending computer's LAN. At the physical layer, also called the PHY layer, the data packets are encoded and modulated onto the carrier medium — a twisted wire pair in the case of a wired network, or electromagnetic radiation in the case of a wireless network — and transmitted to the device with the MAC address resolved at Layer 2.

Transmission of the message across the Internet is achieved through a number of device-to-device hops involving the PHY and Data Link layers of each routing or relaying device in the chain. At each step, the Data Link layer of the receiving device determines the MAC address of the next immediate destination, and the PHY layer transmits the packet to the device with that MAC address.

On arrival at the receiving computer, the PHY layer will demodulate and decode the voltages and frequencies detected from the transmission medium, and pass the received data stream up to the Data Link layer. Here the MAC and LLC elements, such as a message integrity check, will be extracted from the data stream and executed, and the message plus instructions passed up the protocol stack. At Layer 4, a protocol such as Transport Control Protocol (TCP), will ensure that all data frames making up the message have been received and will provide error recovery if any frames have gone missing. Finally the e-mail application will receive the decoded ASCII characters that make up the original transmitted message.

Standards for many layers of the OSI model have been produced by various organisations such as the Institute of Electrical and Electronics Engineers (IEEE). Each standard details the services that are provided within the relevant layer and the protocols or rules that must be followed to enable devices or other layers to call on those services. In fact, multiple standards are often developed for each layer, and they either compete until one emerges as the industry "standard" or else they peacefully coexist in niche areas.

The logical architecture of a wireless network is determined principally by standards that cover the Data Link (LLC plus MAC) and PHY layers of the OSI model. The following sections will give a preliminary introduction to these standards and protocols, while more detailed descriptions will be found in Parts III to V where Local Area (LAN), Personal Area (PAN) and Metropolitan Area (MAN) wireless networking technologies are described respectively.

The next section starts this introductory sketch one layer higher — at the Network layer — not because this layer is specific to wireless networking, but because of the fundamental importance of its addressing and routing functions and of the underlying Internet Protocol (IP).

Network Layer Technologies

The Internet Protocol (IP) is responsible for addressing and routing each data packet within a session or connection set up under the control of transport layer protocols such as TCP or UDP (see Glossary). The heart of the Internet Protocol is the IP address, a 32-bit number that is attached to each data packet and is used by routing software in the network or Internet to establish the source and destination of each packet.

While IP addresses, which are defined at the Network layer, link the billions of devices connected to the Internet into a single virtual network, the actual transmission of data frames between devices relies on the MAC addresses of the network interface cards (NICs), rather than the logical IP addresses of each NIC's host device. Translation between the Layer 3 IP address and the Layer 2 MAC address is achieved using Address Resolution Protocol (ARP), which is described in the Section "Address Resolution Protocol, p. 16".

IP Addressing

The 32-bit IP address is usually presented in "dot decimal" format as a series of four decimal numbers between 0 and 255, for example; 200.100.50.10. This could be expanded in full binary format as 11001000.01100100.00110010.00001010.

As well as identifying a computer or other networked device, the IP address also uniquely identifies the network that the device is connected to. These two parts of the IP address are known as the host ID and the network ID. The network ID is important because it allows a device transmitting a data packet to know what the first port of call needs to be in the route to the packet's destination.

If a device determines that the network ID of the packet's destination is the same as its own network ID, then the packet does not need to be externally routed, for example through the network's gateway and out onto the Internet. The destination device is on its own network and is said to be "local" (Table 2-2). On the other hand, if the destination network ID is different from its own, the destination is a remote IP address and the packet will need to be routed onto the Internet or via some other network bridge to reach its destination. The first stage in this will be to address the packet to the network's gateway.

This process uses two more 32-bit numbers, the "subnet mask" and the "default gateway". A device determines the network ID for a data packet destination by doing a "logical AND" operation on the packet's destination IP address and its own subnet mask. The device determines its own network ID by doing the same operation using its own IP address and subnet mask.

Table 2-2: Local and Remote IP Addresses

Sending Device		
IP Address: Subnet Mask:	200.100.50.10 255.255.255.240	11001000.01100100.00110010.00001010 11111111.11111111.11111111.11110000
Network ID:	200.100.50.000	11001000.01100100.00110010.00000000
Local IP address		
IP Address: Subnet Mask:	200.100.50.14 255.255.255.240	11001000.01100100.00110010.00001110 11111111.11111111.11111111.11110000
Network ID:	200.100.50.000	11001000.01100100.00110010.00000000
Remote IP address		
IP Address: Subnet Mask:	200.100.50.18 255.255.255.240	11001000.01100100.00110010.00010010 11111111.11111111.11111111.11110000
Network ID:	200.100.50.016	11001000.01100100.00110010.00010000

Private IP Addresses

In February 1996, the Network Working Group requested industry comments on RFC 1918, which proposed three sets of so-called private IP addresses (Table 2-3) for use within networks that did not require Internet connectivity. These private addresses were intended to conserve IP address space by enabling many organisations to reuse the same sets of addresses within their private networks. In this situation it did not matter that a computer had an IP address that was not globally unique, provided that that computer did not need to communicate via the Internet.

Table 2-3: Private IP Address Ranges

Class	*Private address range start*	*Private address range end*
A	10.0.0.0	10.255.255.255
B	172.16.0.0	172.31.255.255
C	192.168.0.0	192.168.255.255

Subsequently, the Internet Assigned Numbers Authority (IANA) reserved addresses 169.254.0.0 to 169.254.255.255 for use in Automatic Private IP Addressing (APIPA). If a computer has its TCP/IP configured to obtain an IP address automatically from a DHCP server, but is unable to locate such a server, then the operating system will automatically assign a private IP address from within this range, enabling the computer to communicate within the private network.

Internet Protocol Version 6 (IPv6)

With 32 bits, a total of 2^{32} or 4.29 billion IP addresses are possible — more than enough one would think for all the computers that the human population could possibly want to interconnect.

However, the famous statements that the world demand for computers would not exceed five machines, probably incorrectly attributed to Tom Watson Sr., chairman of IBM in 1943, or the statement of Ken Olsen, founder of Digital Equipment Corporation (DEC), to the 1977 World Future Society convention that "there is no reason for any individual to have a computer in his home", remind us how difficult it is to predict the growth and diversity of computer applications and usage.

The industry is now working on IP version 6, which will give 128-bit IP addresses based on the thinking that a world population of 10 billion by 2020 will eventually be served by many more than one computer each. IPv6 will give a comfortable margin for future growth, with 3.4×10^{38} possible addresses — that is, 3.4×10^{27} for each of the 10 billion population, or 6.6×10^{23} per square metre of the earth's surface.

It seems doubtful that there will ever be a need for IPv7, although, to avoid the risk of joining the short list of famously mistaken predictions of trends in computer usage, it may be as well to add the caveat "on this planet".

Address Resolution Protocol

As noted above, each PHY layer data transmission is addressed to the (Layer 2) MAC address of the network interface card of the receiving device, rather than to its (Layer 3) IP address. In order to address a data packet, the sender first needs to find the MAC address that corresponds to the immediate destination IP address and label the data packet with this MAC address. This is done using Address Resolution Protocol (ARP).

Conceptually, the sending device broadcasts a message on the network that requests the device with a certain IP address to respond with its MAC address. The TCP/IP software operating in the destination device replies with the requested address and the packet can be addressed and passed on to the sender's Data Link layer.

In practice, the sending device keeps a record of the MAC addresses of devices it has recently communicated with, so it does not need to broadcast a request each time. This ARP table or "cache" is looked at first and a broadcast request is only made if the destination IP address is not in the table. In many cases, a computer will be sending the packet to its default gateway and will find the gateway's MAC address from its ARP table.

Routing

Routing is the mechanism that enables a data packet to find its way to a destination, whether that is a device in the next room or on the other side of the world.

A router compares the destination address of each data packet it receives with a table of addresses held in memory — the router table. If it finds a match in the table, it forwards the packet to the address associated with that table entry, which may be the address of another network or of a "next-hop" router that will pass the packet along towards its final destination.

If the router can't find a match, it goes through the table again looking at just the network ID part of the address (extracted using the subnet mask as described above). If a match is found, the packet is sent to the associated address or, if not, the router looks for a default next-hop address and sends the packet there. As a final resort, if no default address is set, the router returns a "Host Unreachable" or "Network Unreachable" message to the sending IP address. When this message is received it usually means that somewhere along the line a router has failed.

What happens if, or when, this elegantly simple structure breaks down? Are there packets out there hopping forever around the Internet, passing from router to router and never finding their destination? The IP header includes a control field that prevents this from happening. The time-to-live (TTL) field is initialised by the sender to a certain value, usually 64, and reduced by one each time the packet passes through a router. When TTL get down to zero, the packet is discarded and the sender is notified using an Internet Control Message Protocol (ICMP) "time-out" message.

Building Router Tables
The clever part of a router's job is building its routing table. For simple networks a static table loaded from a start-up file is adequate but, more generally, Dynamic Routing enables tables to be built up by routers sending and receiving broadcast messages.

These can be either ICMP Router Solicitation and Router Advertisement messages which allow neighbouring routers to ask "Who's there?" and respond "I'm here", or more useful RIP (Router Information Protocol) messages, in which a router periodically broadcasts its complete router table onto the network.

Other RIP and ICMP messages allow routers to discover the shortest path to an address, to update their tables if another router spots an inefficient routing and to periodically update routes in response to network availability and traffic conditions.

A major routing challenge occurs in mesh or mobile ad-hoc networks (MANETs), where the network topology may be continuously changing. One approach to routing in MANETs, inspired by ant behaviour, is described in the Section "Wireless Mesh Network Routing, p. 345".

Network Address Translation

As described in the Section "Private IP Address, p. 15", RFC 1918 defined three sets of private IP addresses for use within networks that do not require Internet connectivity.

However, with the proliferation of the Internet and the growing need for computers in these previously private networks to go online, the limitation of this solution to conserving IP addresses soon became apparent. How could a computer with a private IP address ever get a response from the Internet, when its IP address would not be recognised by any router out in the Internet as a valid destination? Network Address Translation (NAT) provides the solution to this problem.

When a computer sends a data packet to an IP address outside a private network, the gateway that connects the private network to the Internet will replace the private IP source address (192.168.0.1 in Table 2-4), by a public IP address (e.g. 205.55.55.1). The receiving server and Internet routers will recognise this as a valid destination address and route the data packet correctly. When the originating gateway receives a returning data packet it will replace the destination address in the data packet with the original private IP address of the initiating computer. This process of private to public IP address translation at the Internet gateway of a private network is known as Network Address Translation.

Table 2-4: Example of a Simple Static NAT Table

Private IP address	*Public IP address*
192.168.0.1	205.55.55.1
192.168.0.2	205.55.55.2
192.168.0.3	205.55.55.3
192.168.0.4	205.55.55.4

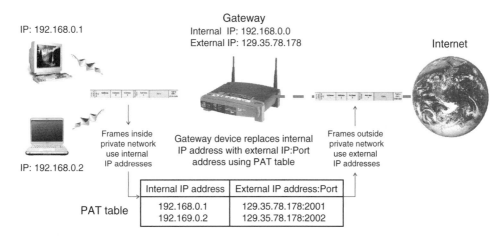

IP: 192.168.0.1

Gateway
Internal IP: 192.168.0.0
External IP: 129.35.78.178

Internet

IP: 192.168.0.2

Frames inside
private network
use internal
IP addresses

Gateway device replaces internal
IP address with external IP:Port
address using PAT table

Frames outside
private network
use external
IP addresses

PAT table

Internal IP address	External IP address:Port
192.168.0.1	129.35.78.178:2001
192.169.0.2	129.35.78.178:2002

Figure 2-2: Port Address Translation in Practice

Static and Dynamic NAT

In practice, similar to routing, NAT can be either static or dynamic. In static NAT, every computer in a private network that requires Internet access has a public IP address assigned to it in a prescribed NAT table. In dynamic NAT, a pool of public IP addresses are available and are mapped to private addresses as required.

Needless to say, dynamic NAT is by far the most common, as it is automatic and requires no intervention or maintenance.

Port Address Translation

One complication arises if the private network's gateway has only a single public IP address available to assign, or if more computers in a private network try to connect than there are IP addresses available to the gateway. This will often be the case for a small organisation with a single Internet connection to an ISP. In this case, it would seem that only one computer within the private network would be able to connect to the Internet at a time. Port Address Translation (PAT) overcomes this limitation by mapping private IP addresses to different port numbers attached to the single public IP address.

When a computer within the private network sends a data packet to be routed to the Internet, the gateway replaces the source address with the single public IP address together with a random port number between 1024 and 65536 (Figure 2-2). When a data packet is returned with this destination

Table 2-5: Example of a Simple PAT Table

Private IP address	*Public IP address:Port*
192.168.0.1	129.35.78.178:2001
192.168.0.2	129.35.78.178:2002
192.168.0.3	129.35.78.178:2003
192.168.0.4	129.35.78.178:2004

address and port number, the PAT table (Table 2-5) enables the gateway to route the data packet to the originating computer in the private network.

Data Link Layer Technologies

The Data Link layer is divided into two sub-layers — Logical Link Control (LLC) and Media Access Control (MAC). From the Data Link layer down, data packets are addressed using MAC addresses to identify the specific physical devices that are the source and destination of packets, rather than the IP addresses, URLs or domain names used by the higher OSI layers.

Logical Link Control

Logical Link Control (LLC) is the upper sub-layer of the Data Link layer (Figure 2-3), and is most commonly defined by the IEEE 802.2 standard. It provides an interface that enables the Network layer to work with any type of Media Access Control layer.

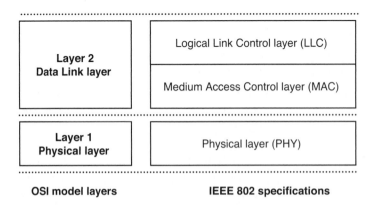

Figure 2-3: OSI Layers and IEEE 802 Specifications

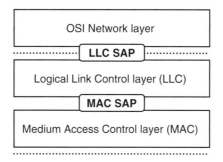

Figure 2-4: Logical Location of LLC and MAC Service Access Points

A frame produced by the LLC and passed down to the MAC layer is called an LLC Protocol Data Unit (LPDU), and the LLC layer manages the transmission of LPDUs between the Link Layer Service Access Points of the source and destination devices. A Link Layer Service Access Point (SAP) is a port or logical connection point to a Network layer protocol (Figure 2-4). In a network supporting multiple Network layer protocols, each will have specific Source SAP (SSAP) and Destination SAP (DSAP) ports. The LPDU includes the 8-bit DSAP and SSAP addresses to ensure that each LPDU is passed on receipt to the correct Network layer protocol.

The LLC layer defines connectionless (Type 1) and connection oriented (Type 2) communication services and, in the latter case, the receiving LLC layer keeps track of the sequence of received LPDUs. If an LPDU is lost in transit or incorrectly received, the destination LLC requests the source to restart the transmission at the last received LPDU.

The LLC passes LPDUs down to the MAC layer at a logical connection point known as the MAC Service Access Point (MAC SAP). The LPDU is then called a MAC Service Data Unit (MSDU) and becomes the data payload for the MAC layer.

Media Access Control

The second sub-layer of the Data Link layer controls how and when a device is allowed to access the PHY layer to transmit data, this is the Media Access Control or MAC layer.

In the following sections, the addressing of data packets at the MAC level is first described. This is followed by a brief look at MAC methods applied in wired networks, which provides an introduction to the more complex solutions required for media access control in wireless networks.

MAC Addressing

A receiving device needs to be able to identify those data packets transmitted on the network medium that are intended for it — this is achieved using MAC addresses. Every network adapter, whether it is an adapter for Ethernet, wireless or some other network technology, is assigned a unique serial number called its MAC address when it is manufactured.

The Ethernet address is the most common form of MAC address and consists of six bytes, usually displayed in hexadecimal, such as 00-D0-59-FE-CD-38. The first three bytes are the manufacturer's code (00-D0-59 in this case is Intel) and the remaining three are the unique serial number of the adapter. The MAC address of a network adapter on a Windows PC can be found in Windows 95, 98 or Me by clicking Start, Run, and then typing "winipcfg", and selecting the adapter, or in Windows NT, 2000, and XP by opening a DOS Window (click Start, Programs, Accessories, Command Prompt) and typing "ipconfig/all".

When an application such as a web browser sends a request for data onto the network, the Application layer request comes down to the MAC SAP as an MSDU. The MSDU is extended with a MAC header that includes the MAC address of the source device's network adapter. When the requested data is transmitted back onto the network, the original source address becomes the new destination address and the network adapter of the original requesting device will detect packets with its MAC address in the header, completing the round trip.

As an example, the overall structure of the IEEE 802.11 MAC frame, or MAC Protocol Data Unit (MPDU) is shown in Figure 2-5.

The elements of the MPDU are as shown in Table 2-6.

Media Access Control in Wired Networks

If two devices transmit at the same time on a network's shared medium, whether wired or wireless, the two signals will interfere and the result will be unusable to both devices. Access to the shared medium therefore needs to be actively managed to ensure that the available bandwidth is not wasted through repeated collisions of this type. This is the main task of the MAC layer.

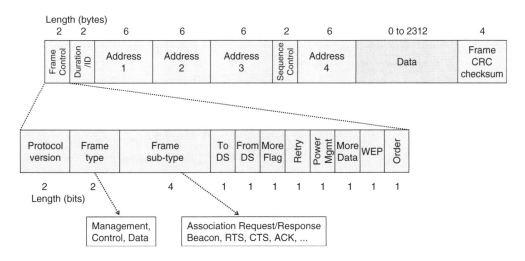

Figure 2-5: MAC Frame Structure

Carrier Sense Multiple Access/Collision Detection (CSMA/CD)

The most commonly used MAC method to control device transmission, and the one specified for Ethernet based networks, is Carrier Sense Multiple Access/Collision Detection (CSMA/CD) (Figure 2-6). When a device has a data frame to transmit onto a network that uses this method, it first checks the physical medium (carrier sensing) to see if any other device is already

Table 2-6: Elements of the 802.11 MPDU Frame Structure

MPDU element	*Description*
Frame control	A sequence of flags to indicate the protocol version (802.11 a/b/g), frame type (management, control, data), sub-frame type (e.g. probe request, authentication, association request, etc.), fragmentation, retries, encryption, etc.
Duration	Expected duration of this transmission. Used by waiting stations to estimate when the medium will again be idle.
Address 1 to Address 4	Destination and source, plus optional to and from addresses within the distribution system.
Sequence	Sequence number to identify frame fragments or duplicates.
Data	The data payload passed down as the MSDU.
Frame check sequence	A CRC-32 checksum to enable transmission errors to be detected.

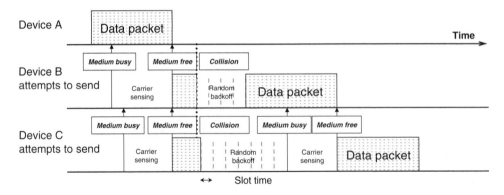

Figure 2-6: Ethernet CSMA/CD Timing

transmitting. If the device senses another transmitting device it waits until the transmission has finished. As soon as the carrier is free it begins to transmit data, while at the same time continuing to listen for other transmissions.

If it detects another device transmitting at the same time (collision detection), it stops transmitting and sends a short jam signal to tell other devices that a collision has occurred. Each of the devices that were trying to transmit then computes a random backoff period within a range 0 to t_{max}, and tries to transmit again when this period has expired. The device that by chance waits the shortest time will be the next to gain access to the medium, and the other devices will sense this transmission and go back into carrier sensing mode.

A very busy medium may result in a device experiencing repeated collisions. When this happens t_{max} is doubled for each new attempt, up to a maximum of 10 doublings, and if the transmission is unsuccessful after 16 attempts the frame is dropped and the device reports an "excessive collision error".

Other Wired Network MAC Methods
Another common form of media access control for wired networks, defined by the IEEE 802.5 standard, involves passing an electronic "token" between devices on the network in a pre-defined sequence. The token is similar to a baton in a relay race in that a device can only transmit when it has captured the token.

If a device does not need control of the media to transmit data it passes the token on immediately to the next device in the sequence, while if it does have data to transmit it can do so once it receives the token. A device can only keep the token and continue to use the media for a specific period of time, after which it has to pass the token on to the next device in the sequence.

Media Access Control in Wireless Networks

The collision detection part of CSMA/CD is only possible if the PHY layer transceiver enables the device to listen to the medium while transmitting. This is possible on a wired network, where invalid voltages resulting from collisions can be detected, but is not possible for a radio transceiver since the transmitted signal would overload any attempt to receive at the same time. In wireless networks such as 802.11, where collision detection is not possible, a variant of CSMA/CD known as CSMA/CA is used, where the CA stands for Collision Avoidance.

Apart from the fact that collisions are not detected by the transmitting device, CSMA/CA has some similarities with CSMA/CD. Devices sense the medium before transmitting and wait if the medium is busy. The duration field in each transmitted frame (see preceding Table 2-6) enables a waiting device to predict how long the medium will be busy.

Once the medium is sensed as being idle, waiting devices compute a random time period, called the contention period, and attempt to transmit after the contention period has expired. This is a similar mechanism to the back-off in CSMA/CD, except that here it is designed to avoid collisions between stations that are waiting for the end of another station's transmitted frame rather than being a mechanism to recover after a detected collision.

CSMA/CA is further described in the Section "The 802.11 MAC Layer, p. 144", where the 802.11 MAC is discussed in more detail, and variations on CSMA/CA used in other types of wireless network will be described as they are encountered.

Physical Layer Technologies

When the MPDU is passed down to the PHY layer, it is processed by the PHY Layer Convergence Procedure (PLCP) and receives a preamble and header, which depend on the specific type of PHY layer in use. The PLCP preamble contains a string of bits that enables a receiver to synchronise its demodulator to the incoming signal timing.

The preamble is terminated by a specific bit sequence that identifies the start of the header, which in turn informs the receiver of the type of modulation and coding scheme to be used to decode the upcoming data unit.

The assembled PLCP Protocol Data Unit (PPDU) is passed to the Physical Medium Dependent (PMD) sublayer, which transmits the PPDU over the physical medium, whether that is twisted-pair, fibre-optic cable, infra-red or radio.

PHY layer technologies determine the maximum data rate that a network can achieve, since this layer defines the way the data stream is coded onto the physical transmission medium. However, the MAC and PLCP headers, preambles and error checks, together with the idle periods associated with collision avoidance or backoff, mean that the PMD layer actually transmits many more bits than are passed down to the MAC SAP by the Data Link layer.

The next sections look at some of the PHY layer technologies applied in wired networks and briefly introduces the key features of wireless PHY technologies.

Physical Layer Technologies — Wired Networks

Most networks that use wireless technology will also have some associated wired networking elements, perhaps an Ethernet link to a wireless access point, a device-to-device FireWire or USB connection, or an ISDN based Internet connection. Some of the most common wired PHY layer technologies are described in this section, as a precursor to the more detailed discussion of local, personal and metropolitan area wireless network PHY layer technologies in Parts III to V.

Ethernet (IEEE 802.3)

The first of these, Ethernet, is a Data Link layer LAN technology first developed by Xerox and defined by the IEEE 802.3 standard. Ethernet uses Carrier Sense Multiple Access with Collision Detection (CSMA/CD), described above, as the media access control method.

Ethernet variants are known as "A" Base-"B" networks, where "A" stands for the speed in Mbps and "B" identifies the type of physical medium used. 10 Base-T is the standard Ethernet, running at 10 Mbps and using an unshielded twisted-pair copper wire (UTP), with a maximum distance of 500 metres between a device and the nearest hub or repeater.

The constant demand for increasing network speed has meant that faster varieties of Ethernet have been progressively developed. 100 Base-T, or Fast Ethernet operates at 100 Mbps and is compatible with 10 Base-T standard Ethernet as it uses the same twisted-pair cabling and CSMA/CD method. The trade-off is with distance between repeaters, a maximum of 205 metres being achievable for 100 Base-T. Fast Ethernet can also use other types of wiring — 100 Base-TX, which is a higher-grade twisted-pair, or 100 Base-FX, which is a two strand fibre-optic cable. Faster speeds to 1 Gbps or 10 Gbps are also available.

The PMD sub-layer is specified separately from the Ethernet standard, and for UTP cabling this is based on the Twisted Pair-Physical Medium Dependent (TP-PMD) specification developed by the ANSI X3T9.5 committee.

The same frame formats and CSMA/CD technology are used in 100 Base-T as in standard 10 Base-T Ethernet, and the 10-fold increase in speed is achieved by increasing the clock speed from 10 MHz to 125 MHz, and reducing the interval between transmitted frames, known as the Inter-Packet Gap (IPG), from 9.6 μs to 0.96 μs. A 125 MHz clock speed is required to deliver a 100 Mbps effective data rate because of the 4B/5B encoding described below.

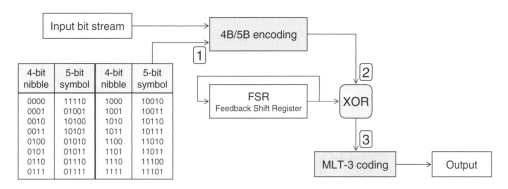

Figure 2-7: 100 Base-T Ethernet Data Encoding Scheme

To overcome the inherent low-pass nature of the UTP physical medium, and to ensure that the level of RF emissions above 30 MHz comply with FCC regulations, the 100 Base-T data encoding scheme was designed to bring the peak power in the transmitted data signal down to 31.25 MHz (close to the FCC limit) and to reduce the power in high frequency harmonics at 62.5 MHz, 125 MHz and above.

4B/5B encoding is the first step in the encoding scheme (Figure 2-7). Each 4-bit nibble of input data has a 5th bit added to ensure there are sufficient transitions in the transmitted bit stream to allow the receiver to synchronise for reliable decoding. In the second step an 11-bit Feedback Shift Register (FSR) produces a repeating pseudo-random bit pattern which is XOR'd with the 4B/5B output data stream. The effect of this pseudo-randomisation is to minimise high frequency harmonics in the final transmitted data signal. The same pseudo-random bit stream is used to recover the input data in a second XOR operation at the receiver.

The final step uses an encoding method called Multi-Level Transition 3 (MLT-3) to shape the transmitted waveform in such a way that the centre frequency of the signal is reduced from 125 MHz to 31.25 MHz.

MLT-3 is based on the repeating pattern 1, 0, −1, 0. As shown in Figure 2-8, an input 1-bit causes the output to transition to the next bit in the pattern while an input 0-bit causes no transition, i.e. the output level remaining unchanged. Compared to the Manchester Phase Encoding (MPE) scheme used in 10 Base-T Ethernet, the cycle length of the output signal is reduced by a factor of 4, giving a signal peak at 31.25 MHz instead of 125 MHz. On the physical UTP medium, the 1, 0 and −1 signal levels are represented by line voltages of +0.85, 0.0 and −0.85 Volts.

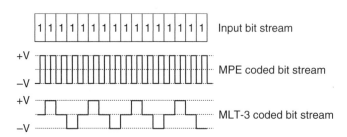

Figure 2-8: Ethernet MPE and Fast Ethernet MLT-3 Encoding

ISDN

ISDN, which stands for Integrated Services Digital Network, allows voice and data to be transmitted simultaneously over a single pair of telephone wires. Early analogue phone networks were inefficient and error prone as a long distance data communication medium and, since the 1960s, have gradually been replaced by packet-based digital switching systems.

The International Telephone and Telegraph Consultative Committee (CCITT), the predecessor of the International Telecommunications Union (ITU), issued initial guidelines for implementing ISDN in 1984, in CCITT Recommendation I.120. However, industry-wide efforts to establish a specific implementation for ISDN only started in the early 1990s when US industry members agreed to create the National ISDN 1 standard (NI-1). This standard, later superseded by National ISDN 2 (NI-2), ensured the interoperability of end user and exchange equipment.

Two basic types of ISDN service are defined — Basic Rate Interface (BRI) and Primary Rate Interface (PRI). ISDN carries voice and user data streams on "bearer" (B) channels, typically occupying a bandwidth of 64 kbps, and control data streams on "demand" (D) channels, with a 16 kbps or 64 kbps bandwidth depending on the service type.

BRI provides two 64 kbps B channels, which can be used to make two simultaneous voice or data connections or can be combined into one 128 kbps connection. While the B channels carry voice and user data transmission, the D channel is used to carry Data Link and Network layer control information.

The higher capacity PRI service provides 23 B channels plus one 64 kbps D channel in the US and Japan, or 30 B channels plus one D channel in Europe. As for BRI, the B channels can be combined to give data bandwidths of 1472 kbps (US) or 1920 kbps (Europe).

As noted above, telephone wires are not ideal as a digital communication medium. The ISDN PHY layer limits the effect of line attenuation, near-end and far-end crosstalk and noise by using Pulse Amplitude Modulation (PAM) technology (see the Section "Pulse Modulation Methods, p. 104") to achieve a high data rate at a reduced transmission rate on the line.

This is achieved by converting multiple (often two or four) binary bits into a single multilevel transmitted symbol. In the US, the 2B1Q method

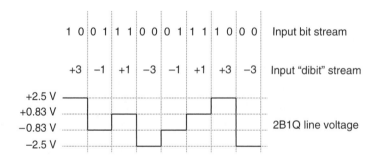

Figure 2-9: 2B1Q Line Code Using in ISDN

is used, which converts two binary bits (2B) into a single output symbol, known as a "quat" (1Q), which can have one of four values (Figure 2-9 and Table 2-7). This effectively halves the transmission rate on the line, so that a 64 kbps data rate can be transmitted at a symbol rate of 32 ksps, achieving higher data rates within the limited bandwidth of the telephone system.

As well as defining a specific PHY layer, ISDN also specifies Data Link and Network layer operation. LAP-D (Link Access Protocol D-channel) is a Data Link protocol, defined in ITU-T Q.920/921, that ensures error free transmission on the PHY layer. Two Network layer protocols are defined in ITU-T Q.930 and ITU-T Q.931 to establish, maintain and terminate user-to-user, circuit-switched, and packet-switched network connections.

FireWire
FireWire, also known as IEEE 1394 or i.Link, was developed by Apple Computer Inc. in the mid-1990s as a local area networking technology. At that time it provided a 100 Mbps data rate, well above the Universal Serial Bus (USB) speed of 12 Mbps, and it was soon taken up

Table 2-7: 2B1Q Line Code Used in ISDN

Input "DIBIT"	Output "QUAT"	Line voltage
10	+3	+2.5
11	+1	+.833
00	−1	−.833
01	−3	−2.5

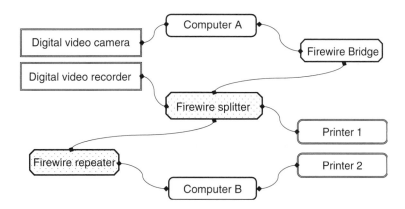

Figure 2-10: FireWire Network Topology: Daisy-chain and Tree Structures

by a number of companies for applications such as connecting storage and optical drives.

FireWire is now supported by many electronics and computer companies, often under the IEEE 1394 banner, because of its ability to reliably and inexpensively transmit digital video data at high speeds, over single cable lengths of up to 4.5 metres. The standard data rate is 400 Mbps, although a faster version is also available delivering 800 Mbps and with plans for 3.2 Gbps. Range can be extended up to 72 metres using signal repeaters in a 16-link daisy chain, and FireWire to fibre transceivers are also available that replace the copper cable by optical fibre and can extend range to 40 km. A generic FireWire topology is shown in Figure 2-10.

The FireWire standard defines a serial input/output port and bus, a 4 or 6 wire dual-shielded copper cable that can carry both data and power, and the related Data Link, Network and Transport layer protocols. FireWire is based on the Control and Status Register Management (CSR) architecture, which means that all interconnected devices appear as a single memory of up to 256 Terabytes (256×10^{12} bytes). Each transmitted packet of data contains three elements: a 10-bit bus ID that is used to determine which FireWire bus the data packet originated from, a 6-bit ID that identifies which device or node on that bus sent the data packet, and a 48-bit offset that is used to address registers and memory in a node.

While primarily used for inter-device communication, The Internet Society has combined IP with the FireWire standard to produce a standard called IP over IEEE 1394, or IP 1394. This makes it possible for networking services such as FTP, HTTP and TCP/IP to run on the high speed FireWire PHY layer as an alternative to Ethernet.

An important feature of FireWire is that the connections are "hot-swappable", which means that a new device can be connected, or an existing device disconnected, while the connection is live. Devices are automatically assigned node IDs, and these IDs can change as the network topology changes. Combining the node ID variability of FireWire with the IP requirement for stable IP addresses of connected devices, presents one of the interesting problems in enabling IP connections over FireWire. This is solved using a special Address Resolution Protocol (ARP) called 1394 ARP.

In order to uniquely identify a device in the network, 1394 ARP uses the 64-bit Extended Unique Identifier (EUI-64), a unique 64-bit number that is assigned to every FireWire device on manufacture. This is an extended version of the MAC address, described in the Section "Media Access Control, p. 21" that is used to address devices other than network interfaces. A 48-bit MAC address can be converted into a 64-bit EUI-64 by prefixing the two hexadecimal octets "FF-FF".

Universal Serial Bus

The Universal Serial Bus (USB) was introduced in the mid-1990s to provide a hot-swappable "plug-and-play" interface that would replace different types of peripheral interfaces (parallel ports, serial ports, PS/2, MIDI, etc.) for devices such as joysticks, scanners, keyboards and printers. The maximum bandwidth of USB 1.0 was 12 Mbps, but this has since increased to a FireWire matching 480 Mbps with USB 2.0.

USB uses a host-centric architecture, with a host controller dealing with the identification and configuration of devices connected either directly to the host or to intermediate hubs (Figure 2-11). The USB specification supports both isochronous and asynchronous transfer types over the same connection. Isochronous transfers require guaranteed bandwidth and low latency for applications such as telephony and media streaming, while asynchronous transfers are delay-tolerant and are able to wait for available bandwidth. USB control protocols are designed specifically to

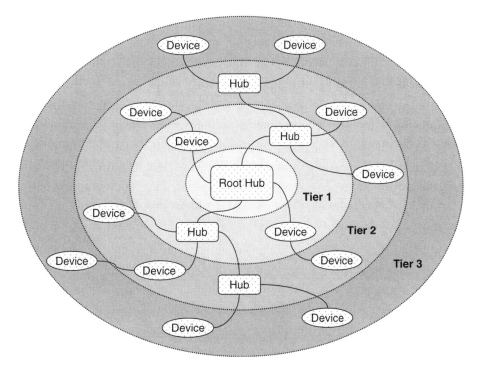

Figure 2-11: USB Network Topology: Daisy-chain and Tree Structures

give a low protocol overhead, resulting in highly effective utilisation of the available bandwidth.

This available bandwidth is shared among all connected devices and is allocated using "pipes", with each pipe representing a connection between the host and a single device. The bandwidth for a pipe is allocated when the pipe is established, and a wide range of different device bit rates and device types can be supported concurrently. For example, digital telephony devices can be concurrently accommodated ranging from 1 "bearer" plus 1 "demand" channel (64 kbps — see ISDN above) up to T1 capacity (1.544 Mbps).

USB employs NRZI (Non Return to Zero Inverted) as a data encoding scheme. In NRZI encoding, a 1-bit is represented by no change in output voltage level and a 0-bit is represented by a change in voltage level (Figure 2-12). A string of 0-bits therefore causes the NRZI output to toggle between states on each bit cycle, while a string of 1-bits causes a period with no transitions in the output.

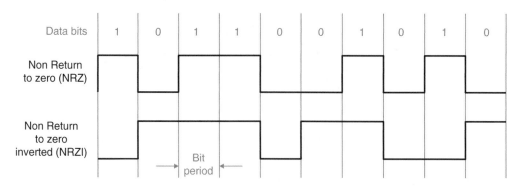

Figure 2-12: USB NRZI Data Encoding Scheme

NRZI has the advantage of a somewhat improved noise immunity compared with the straight encoding of the input data stream as output voltages.

Physical Layer Technologies — Wireless Networks

The PHY layer technologies that provide the Layer 1 foundation for wireless networks will be described further in Parts III, IV and V, where LAN, PAN and MAN technologies and their implementations will be covered in detail.

Each wireless PHY technology, from Bluetooth to ZigBee, will be described in terms of a number of key aspects, as summarised in Table 2-8.

The range and significance of the issues vary depending on the type of technology (Ir, RF, Near-field) and its application (PAN, LAN or WAN).

Operating System Considerations

In order to support networking, an operating system needs as a minimum to implement networking protocols, such as TCP/IP, and the device drivers required for network hardware. Early PC operating systems, including Windows versions prior to Windows 95, were not designed to support networking. However, with the rise of the Internet and other networking technologies, virtually every operating system today qualifies as a network operating system (NOS).

Individual network operating systems offer additional networking features such as firewalls, simplified set-up and diagnostic tools, remote access,

Table 2-8: Aspects of PHY and Data Link Layer Wireless Technologies

Technology aspect	*Issues and considerations*
Spectrum	What part of the electromagnetic spectrum is used, what is the overall bandwidth available, how is this segmented into channels? What mechanisms are available to control utilised bandwidth to ensure coexistence with other users of the same spectrum?
Propagation	What power levels are permitted by regulatory authorities in the spectrum in question? What mechanisms are available to control the transmitted power or propagation pattern to minimise co-channel interference for other users, maximise effective range or utilise spatial diversity to increase throughput?
Modulation	How is encoded data carried on the physical medium, for example by modulating one or more carriers in phase and/or amplitude, or by modulating pulses in amplitude and/or position?
Data encoding	How are the raw bits of a data frame coded into symbols for transmission? What functions do these coding mechanisms serve, for example increasing robustness to noise or increasing the efficient use of available bandwidth?
Media access	How is access to the transmission medium controlled to ensure that the bandwidth available for data transmission is maximised and that contention between users is efficiently resolved? What mechanisms are available to differentiate media access for users with differing service requirements?

and inter-connection with networks running other operating systems, as well as support for network administration tasks such as enforcing common settings for groups of users.

The choice of a network operating system will not be covered in detail here, but should be based on a similar process to that described in Parts III and IV for selecting WLAN and WPAN technologies. Start by determining networking service requirements such as security, file sharing, printing and messaging. The two main network operating systems are the Microsoft Windows and Novell NetWare suites of products. A key differentiator between these two products may be a requirement for interoperability support in networks that include other operating systems such as UNIX or Linux. NetWare is often the preferred NOS in mixed operating-system networks, while simplicity of installation and administration makes Windows the preferred product suite in small networks where technical support may be limited.

Summary

The OSI network model provides the conceptual framework to describe the logical operation of all types of networks, from a wireless PAN link between a mobile phone and headset to the global operation of the Internet.

The key features that distinguish different networking technologies, particularly wired and wireless, are defined at the Data Link (LLC and MAC) and physical (PHY) layers. These features will be covered in detail in Parts III to V which, above all, reveal the fascinating variety of different techniques that have been harnessed to bring wireless networks to life.

CHAPTER 3

Wireless Network Physical Architecture

Wired Network Topologies — A Refresher

The topology of a wired network refers to the physical configuration of links between networked devices or nodes, where each node may be a computer, an end-user device such as a printer or scanner, or some other piece of network hardware such as a hub, switch or router.

The building block from which different topologies are constructed is the simple point-to-point wired link between two nodes, shown in Figure 3-1. Repeating this element results in the two simplest topologies for wired networks — bus and ring.

For the ring topology, there are two possible variants depending on whether the inter-node links are simplex (one-way) or duplex (two-way). In the simplex case, each inter-node link has a transmitter at one end and a receiver at the other, and messages circulate in one direction around the ring, while in the duplex case each link has both transmitter and receiver (a so-called transceiver) at each end, and messages can circulate in either direction.

Bus and ring topologies are susceptible to single-point failures, where a single broken link can isolate sections of a bus network or halt all traffic in the case of a ring.

The step that opens up new possibilities is the introduction of specialised network hardware nodes designed to control the flow of data between

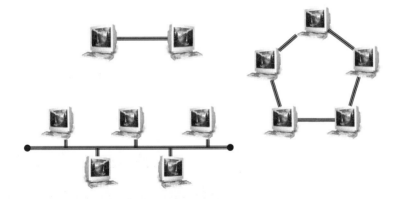

Figure 3-1: Point-to-point, Bus and Ring Topologies

other networked devices. The simplest of these is the passive hub, which is the central connection point for LAN cabling in star and tree topologies, as shown in Figure 3-2. An active hub, also known as a repeater, is a variety of passive hub that also amplifies the data signal to improve signal strength over long network connections.

For some PAN technologies, such as USB, star and tree topologies can be built without the need for specialised hardware,

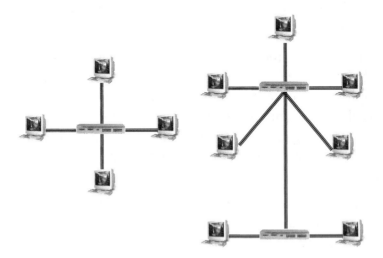

Figure 3-2: Star and Tree Topologies

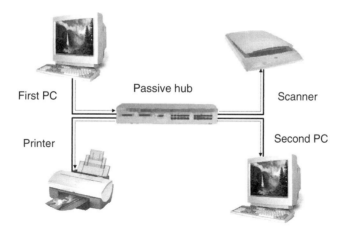

Figure 3-3: A Passive Hub in a Physical Star Network

because of the daisy-chaining capability of individual devices (see Figure 2-11).

An active or passive hub in a star topology LAN transmits every received data packet to every connected device. Each device checks every packet and decodes those identified by the device's MAC address. The disadvantage of this arrangement is that the bandwidth of the network is shared among all devices, as shown in Figure 3-3. For example, if two PCs are connected through a 10 Mbps passive hub, each will have on average 5 Mbps of bandwidth available to it.

If the first PC is transmitting data, the hub relays the data packets on to all other devices in the network. Any other device on the network will have to wait its turn to transmit data.

A switching hub (or simply a switch) overcomes this bandwidth sharing limitation by only transmitting a data packet to the device to which it is addressed. Compared to a non-switching hub, this requires increased memory and processing capability, but results in a significant improvement in network capacity.

The first PC (Figure 3-4) is transmitting data stream A to the printer and the switch directs these data packets only to the addressed device. At the same time, the scanner is sending data stream B to the second PC.

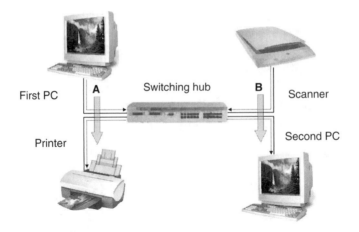

Figure 3-4: Switching Hub in Physical Star Network.

The switch is able to process both data stream concurrently, so that the full network bandwidth is available to every device.

Wireless Network Topologies

Point to Point Connections

The simple point to point connection shown in Figure 3-1 is probably more common in wireless than in wired networks, since it can be found in a wide variety of different wireless situations, such as:

- peer-to-peer or ad-hoc Wi-Fi connections
- wireless MAN back-haul provision
- LAN wireless bridging
- Bluetooth
- IrDA

Star Topologies in Wireless Networks

In wireless networks the node at the centre of a star topology (Figure 3-5), whether it is a WiMAX base station, Wi-Fi access point, Bluetooth Master device or a ZigBee PAN coordinator, plays a similar role to the hub in a wired network. As described in Parts III to V, the different wireless

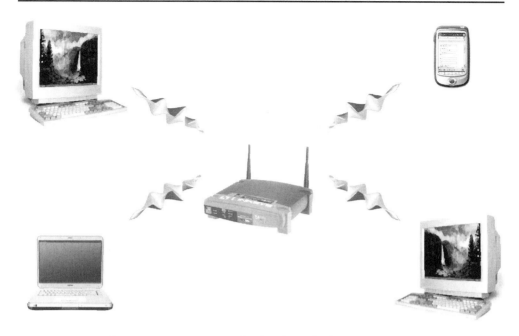

Figure 3-5: Star Topologies in Wireless Networks

networking technologies require and enable a wide range of different functions to be performed by these central control nodes.

The fundamentally different nature of the wireless medium means that the distinction between switching and non-switching hubs is generally not relevant for control nodes in wireless networks, since there is no direct wireless equivalent of a separate wire to each device. The wireless LAN switch or controller (Figure 3-6), described in the Section "Wireless LAN Switches or Controllers, p. 48", is a wired network device that switches data to the access point that is serving the addressed destination station of each packet.

The exception to this general rule arises when base stations or access point devices are able to spatially separate individual stations or groups of stations using sector or array antennas. Figure 3-7 shows a wireless MAN example, with a switch serving four base station transmitters each using a 90° sector antenna.

With this configuration, the overall wireless MAN throughput is multiplied by the number of transmitters, similar to the case of the wired switching hub shown in Figure 3-4.

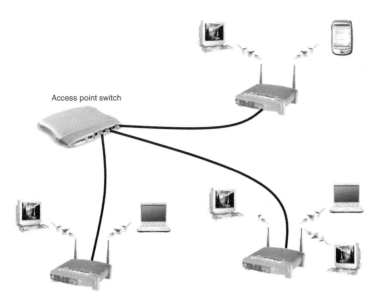

Access point switch

Figure 3-6: A Tree Topology Using a Wireless Access Point Switch

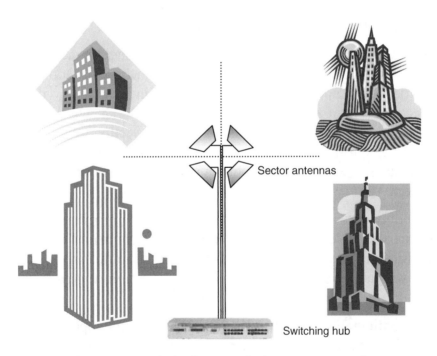

Sector antennas

Switching hub

Figure 3-7: Switched Star Wireless MAN Topology

In the wireless LAN case, a similar spatial separation can be achieved using a new class of device called an access point array, described below in the Section "Wireless LAN Arrays, p. 52", which combines a wireless LAN controller with an array of sector antennas to multiply network capacity. The general technique of multiplying network throughput by addressing separate spatial zones or propagation paths is known as space division multiplexing (Section "Space Division Multiple Access, p. 94"), and finds its most remarkable application in MIMO radio, described in the Section "MIMO Radio, p. 124".

Mesh Networks

Mesh networks, also known as mobile ad hoc networks (MANETs), are local or metropolitan area networks in which nodes are mobile and communicate directly with adjacent nodes without the need for central controlling devices. The topology of a mesh, shown generically in Figure 3-8, can be constantly changing, as nodes enter and leave the network, and data packets are forwarded from node-to-node towards their destination in a process called hopping.

The data routing function is distributed throughout the entire mesh rather than being under the control of one or more dedicated devices. This is similar to the way that data travels around the Internet, with a packet hopping from one device to another until it reaches its destination, although in mesh networks, the routing capabilities are included in every node rather than just in dedicated routers.

This dynamic routing capability requires each device to communicate its routing information to every device it connects with, and to update this as nodes move within, join and leave the mesh.

This distributed control and continuous reconfiguration allows for rapid re-routing around overloaded, unreliable or broken paths, allowing mesh networks to be self-healing and very reliable, provided that the density of nodes is sufficiently high to allow alternative paths. A key challenge in the design of the routing protocol is to achieve this continuous re-configuration capability with a manageable overhead in terms of data bandwidth taken up by routing information messages. One approach to this problem, the biologically inspired AntHocNet, is described in the Section "Wireless Mesh Network Routing, p. 345".

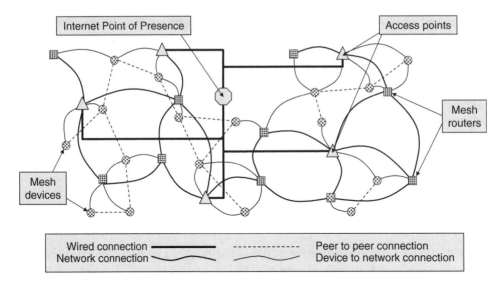

Figure 3-8: Mesh Network Topology

The multiplicity of paths in a mesh network has a similar impact on total network throughput as the multiple paths shown in Figures 3-4 and 3-7 for the case of wired network switches and sectorised wireless networks. Mesh network capacity will grow as the number of nodes, and therefore the number of usable alternative paths, grow, so that capacity can be increased simply by adding more nodes to the mesh.

As well as the problem of efficiently gathering and updating routing information, mesh networks face several additional technical challenges such as:

- Wireless link reliability — a packet error rate that may be tolerable over a single hop in an hub and spokes configuration will quickly compound over multiple hops, limiting the size to which a mesh can grow and remain effective.

- Seamless roaming — seamless connection and reconnection of moving nodes has not been a requirement in most wireless network standards, although 802.11 Task Groups TGr and TGs are addressing this.

- Security — how to authenticate users in a network with no stable infrastructure?

From a practical standpoint, the self-configuring, self-optimising and self-healing characteristics of mesh networks eliminate many of the management and maintenance tasks associated with large-scale wireless network deployments.

ZigBee (Section "ZigBee (IEEE 802.15.4), p. 273") is one standard which explicitly supports mesh networks and the IEEE 802.11 Task Group TGs is in the process of developing a standard which addresses WLAN mesh networks. Two industry bodies have already been established to promote 802.11s mesh proposals, the Wi-Mesh Alliance and SEEMesh (Simple, Efficient and Extensible Mesh).

Wireless LAN Devices

Wireless Network Interface Cards

The wireless network interface card (NIC) turns a device such as a PDA, laptop or desktop computer into a wireless station and enables the device to communicate with other stations in a peer-to-peer network or with an access point.

Wireless NICs are available in a variety of form factors (Figure 3-9), including PC (Type II PCMCIA) and PCI cards, as well as external USB devices and USB dongles, or compact flash for PDAs. Most wireless NICs

Figure 3-9: A Variety of Wireless NIC Forms (courtesy of Belkin Corporation, D-Link (Europe) Ltd. and Linksys (a division of Cisco Systems Inc.))

have integrated antennas, but a few manufacturers provide NICs with an external antenna connection or detachable integrated antenna which can be useful to attach a high-gain antenna when operating close to the limit of wireless range.

There are few features to distinguish one wireless NIC from another. Maximum transmitter power is limited by local regulatory requirements and, for standards based equipment, certification by the relevant body (such as Wi-Fi certification for 802.11) will ensure interoperability of equipment from different manufacturers. The exception will be proprietary extensions or equipment released prior to standard ratification, such as "pre-n" hardware announced by some manufacturers in advance of 802.11n ratification.

High-end mobile products, particularly laptop computers, are increasingly being shipped with integrated wireless NICs, and with Intel's Centrino® technology the wireless LAN interface became part of the core chipset family.

Access Points

The access point (AP) is the central device in a wireless local area network (WLAN) that provides the hub for wireless communication with the other stations in the network. The access point is usually connected to a wired network and provides a bridge between wired and wireless devices.

The first generation of access points, now termed "fat" access points, began to appear after the ratification of the IEEE 802.11b standard in 1999, and provided a full range of processing and control functions within each unit, including:

- security features, such as authentication and encryption support
- access control based on lists or filters
- SNMP configuration capabilities

Transmit power level setting, RF channel selection, security encryption and other configurable parameters required user configuration of the access point, typically using a web-based interface.

As well as providing this basic functionality, access points designed for home or small office wireless networking typically include a number of additional networking features, as shown in Table 3-1.

Table 3-1: Optional Access Point Functionality

Feature	*Description*
Internet gateway	Supporting a range of functions such as: routing, Network Address Translation, DHCP server providing dynamic IP addresses to client stations, and Virtual Private Network (VPN) passthrough.
Switching hub	Several wired Ethernet ports may be included that provide switching hub capabilities for a number of Ethernet devices.
Wireless bridge or repeater	Access point that can function as a relay station, to extend the operating range of another access point, or as a point-to-point wireless bridge between two networks.
Network storage server	Internal hard drives or ports to connect external storage, providing centralised file storage and back-up for wireless stations.

Figure 3-10 illustrates a range of access point types, including weatherproofed equipment for outdoor coverage.

In contrast to the first generation "fat" access point described above, slimmed-down "thin" access points are also available that limit access point capabilities to the essential RF communication functions and rely on the centralisation of control functions in a wireless LAN switch.

Figure 3-10: First Generation Wireless Access Points (courtesy of Belkin Corporation, D-Link (Europe) Ltd. and Linksys (a division of Cisco Systems Inc.))

Wireless LAN Switches or Controllers

In a large wireless network, typically in a corporate environment with tens and perhaps hundreds of access points, the need to individually configure access points can make WLAN management a complicated task. Wireless LAN switches simplify the deployment and management of large-scale WLANs. A wireless LAN switch (also known as a wireless LAN controller or access router), is a networking infrastructure device designed to handle a variety of functions on behalf of a number of dependent, or "thin", access points (Figure 3-11).

As shown in Table 3-2, this offers several advantages for large-scale WLAN implementations, particularly those supporting voice services.

The driver behind the development of the wireless switch is to enable the task of network configuration and management, which becomes increasingly complex and time consuming as wireless networks grow. A wireless switch provides centralised control of configuration, security, performance monitoring and troubleshooting, which is essential in an enterprise scale wireless LAN.

Taking security as an example, with WEP, WPA, and 802.11i, all potentially in use at the same time in a large WLAN deployment, if security

Table 3-2: "Thin" Access Point Advantages

Advantage	*Description*
Lower cost	A "thin" access point is optimised to cost effectively implement wireless communication functions only, reducing initial hardware cost as well as future maintenance and upgrade costs.
Simplified access point management	Access point configuration, including security functions, is centralised in order to simplify the network management task.
Improved roaming performance	Roaming handoffs are much faster than with conventional access points, which improves the performance of voice services.
Simplified network upgrades	The centralised command and control capability makes it easier to upgrade the network in response to evolving WLAN standards, since upgrades only have to be applied at the switch level, and not to individual access points.

configuration has to be managed for individual access points, the routine management of encryption keys and periodic upgrade of security standards for each installed access point quickly becomes unmanageable. With a centralised security architecture provided by a wireless switch, these management tasks only need to be completed once.

WLAN switches also provide a range of additional features, not found in first generation access points, as described in Table 3-3.

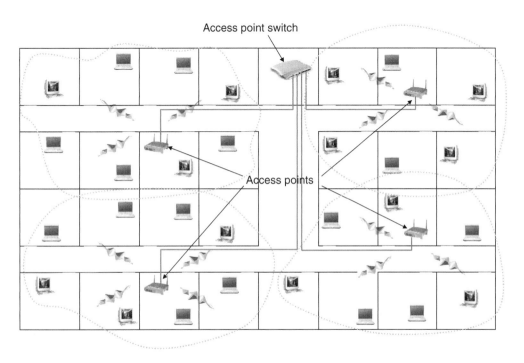

Figure 3-11: Wireless LAN Topology Using a Wireless Switch

Lightweight Access Point Protocol
Centralising command and control into a wireless LAN switch device introduces the need for a communication protocol between the switch and its dependent access points, and the need for interoperability requires that this protocol is based on an industry standard.

The Lightweight Access Point Protocol (LWAPP) standardises communications between switches or other hub devices and access points, and was initially developed by the Internet Engineering Task Force (IETF).

Table 3-3: Wireless LAN Switch Features

Feature	*Description*
Layout planning	Automated site survey tools that allow import of building blueprints and construction specifications and determine optimal access point locations.
RF management	Analysis of management frames received from all access points enables RF signal related problems to be diagnosed and automatically corrected, by adjusting transmit power level or channel setting of one or more access points.
Automatic configuration	Wireless switches can provide automatic configuration by determining the best RF channel and transmit power settings for individual access points.
Load balancing	Maximising network capacity by automatic load balancing of users across multiple access points.
Policy-based access control	Access policies can be based on access point groupings and client lists that specify which access points or groups specific client stations are permitted to connect to.
Intrusion detection	Rogue access points and unauthorised users or ad hoc networks can be detected and located, either by continuous scanning or by scheduled site surveys.

The IETF specification describes the goals of the LWAPP protocol as follows:

■ To reduce the amount of protocol code being executed at the access point so that efficient use can be made of the access point's computing power, by applying this to wireless communication rather than to bridging, forwarding and other functions.

■ To use centralised network computing power to execute the bridging, forwarding, authentication, encryption and policy enforcement functions for a wireless LAN.

■ To provide a generic encapsulation and transport mechanism for transporting frames between hub devices and access points, which will enable multi-vendor interoperability and ensure that LWAPP can be applied to other access protocols in the future.

Table 3-4: LWAPP Functions

LWAPP function	*Description*
Access point device discovery and information exchange	An access point sends a Discovery Request frame and any receiving access router responds with a Discovery Reply frame. The access point selects a responding access router and associates by exchanging Join Request and Join Reply frames.
Access point certification, configuration, provisioning and software control	After association, the access router will provision the access point, providing a Service Set Identifier (SSID), security parameters, operating channel and data rates to be advertised. The access router can also configure MAC operating parameters (e.g. number of transmission attempts for a frame), transmit power, DSSS or OFDM parameters and antenna configuration in the access point. After provisioning and configuration, the access point is enabled for operation.
Data and management frame encapsulation, fragmentation and formatting	LWAPP encapsulates data and management frames for transport between the access point and access router. Fragmentation of frames and re-assembly of fragment will be handled if the encapsulated data or management frames exceed the Maximum Transmission Unit (MTU) supported between the access point and access router.
Communication control and management between access points and associated devices	LWAPP enables the access router to request statistical reports from its access points, including data about the communication between the access point and its associated devices (e.g. retry counts, RTS/ACK failure counts).

The main communication and control functions that are achieved using LWAPP are summarised in Table 3-4.

Although the initial draft specification for LWAPP expired in March 2004, a new IETF working group called Control and Provisioning of Wireless Access Points (CAPWAP) was formed, with most working group members continuing to recommend LWAPP over alternatives such as Secure Light Access Point Protocol (SLAPP), Wireless LAN Control Protocol (WICOP) and CAPWAP Tunnelling Protocol (CTP). It seems likely that LWAPP will be the basis of an eventual CAPWAP protocol.

Wireless LAN Arrays

The so-called "3rd generation" architecture for WLAN deployment uses a device called an access point array, which is the LAN equivalent of the sectorised WMAN base station illustrated in Figure 3-7.

A single access point array incorporates a wireless LAN controller together with 4, 8 or 16 access points, which may combine both 802.11a and 802.11b/g radio interfaces. A typical example uses 4 access points for 802.11a/g coverage, employing 180° sector antennas offset by 90°, and 12 access points for 802.11a covering, with 60° sector antennas offset by 30°, as illustrated in Figure 3-12.

This type of device, with 16 access points operating 802.11a and g networks at an individual headline data rate of 54 Mbps, offers a total wireless LAN capacity of 864 Mbps. The increased gain of the sector antennas also means that the operating range of an access point array can be double or more the range of a single access point with an omnidirectional antenna.

For high capacity coverage over a larger operating area, multiple access point arrays, controlled by a second tier of WLAN controllers would

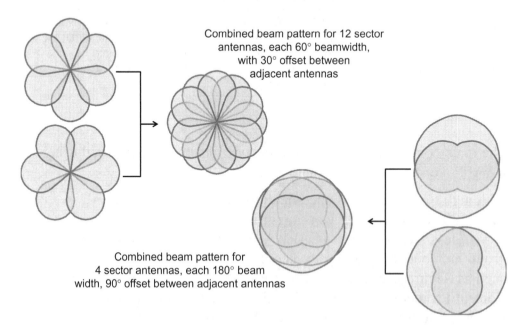

Combined beam pattern for 12 sector antennas, each 60° beamwidth, with 30° offset between adjacent antennas

Combined beam pattern for 4 sector antennas, each 180° beam width, 90° offset between adjacent antennas

Figure 3-12: Antenna Configuration in a 16-sector Access Point Array

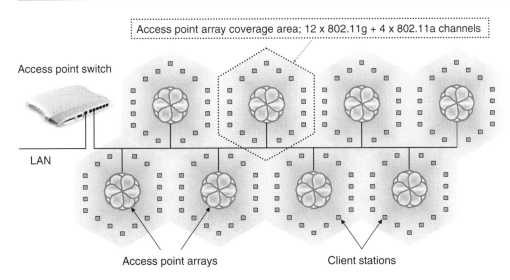

Figure 3-13: WLAN Tree Topology Employing Access Point Arrays

create a tree topology, as shown in Figure 3-13, with multi giga-bit total WLAN capacity.

Miscellaneous Wireless LAN Hardware

Wireless Network Bridging

Wireless bridge components that provide point-to-point WLAN or WMAN links are available from a number of manufacturers, packaged in weather proof enclosures for outdoor use (Figure 3-14). The D-Link DWL 1800 is one example which bundles a 16 dBi flat panel antenna with a 2.4 GHz radio providing a transmit power of 24 dBm (under FCC) or 14 dBm (under ETSI regulations), to deliver a range of 25 km under FCC or 10 km under ETSI.

Many simple wireless LAN access points also support network bridging, or can be upgraded with a firmware upgrade to provide this capability. Configuring these devices simply involves entering the MAC address of the other endpoint into each station's access control list, so that each station only decodes packets transmitted by the other endpoint of the bridge.

Wireless Printer Servers

A wireless printer server allows a printer to be flexibly shared among a group of users in the home or office without the need for the printer to be hosted by one computer or to be connected to a wired network.

Figure 3-14: Outdoor Wireless Bridges (courtesy of D-Link (Europe) Ltd. and Linksys (a division of Cisco Systems Inc.))

Typically, as well as wired Ethernet and wireless LAN interfaces, this device may include one or more different types of printer connections, such as USB or parallel printer ports, as well as multiple ports to enable multiple printers to be connected — such as a high-speed black and white laser and a separate colour printer.

Figure 3-15: Wireless Printer Servers (courtesy of Belkin Corporation, D-Link (Europe) Ltd. and Linksys (a division of Cisco Systems Inc.))

A printer server for home or small office wireless networking may also be bundled with a 4-port switch to enable other wired network devices to share the printer and use the wireless station as a bridge to other devices on the wireless network. Figure 3-15 shows a range of wireless printer servers.

Wireless LAN Antennas

Traditional Fixed Gain Antennas

Antennas for 802.11b and 11g networks, operating in the 2.4 GHz ISM band, are available to achieve a variety of coverage patterns. The key features that dictate the choice of antenna for a particular application are gain, measured in dBi (see the Section "Antenna Gain, p. 107") and angular beamwidth, measured in degrees.

The most common WLAN antenna, standard in all NICs and in most access points, is the omnidirectional antenna, which has a gain in the range from 0 to 7 dBi and a beamwidth, perpendicular to the antenna axis, of a full 360°. A range of WLAN antennas is shown in Figure 3-16 and typical parameters are summarised in Table 3-5.

For sector antennas with a given horizontal beamwidth, the trade-off for higher gain is a narrower vertical beamwidth, which will result in a smaller coverage area at a given distance and will require more precise alignment.

A further important feature of an antenna is its polarisation, which refers to the orientation of the electric field in the electromagnetic wave emitted

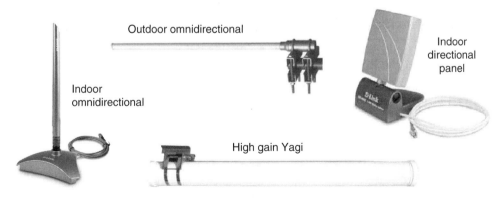

Figure 3-16: Wireless LAN Antenna Types (courtesy of D-Link (Europe) Ltd.)

Table 3-5: Typical Wireless LAN Antenna Parameters for 2.4 GHz Operation

Antenna type	*Sub-type*	*Beamwidth (Degrees)*	*Gain (dBi)*
Omnidirectional		360	0–15
Patch / Panel		15–75	8–20
Sector		180	8–15
		120	9–20
		90	9–20
		60	10–17
Directional	Yagi	10–30	8–20
	Parabolic reflector	5–25	14–30

by the antenna. Most common antennas, including all those listed in the table above, produce linearly polarised waves, with the electrical field oriented either vertically or horizontally — hence vertical or horizontal polarisation. WLAN antennas that produce circular polarisation are also available (helical antennas) but are less common.

It is important that the polarisations of transmitting and receiving antennas are matched, since a vertical polarised receiving antenna will be unable to receive a signal transmitted by a horizontally polarised transmitting antenna, and vice versa. It is equally important for antennas to be correctly mounted, as rotating an antenna 90° about the direction of propagation will change its polarisation by the same angle (e.g. from horizontal to vertical).

Although WLAN operation in the 5 GHz band has developed more recently than in the 2.4 GHz band, a similar selection of antennas is available for the higher-frequency band. A variety of dual band omnidirectional and patch antennas are also available to operate in both WLAN bands.

Smart Antennas

The data throughput of a wireless network that uses a traditional antenna of the type described above is limited because only one network node at a time can use the medium to transmit a data packet. (Other so-called

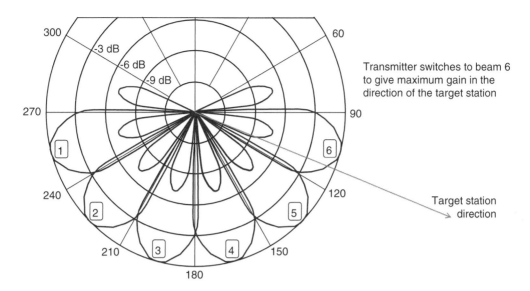

Figure 3-17: Beam Pattern of a Six Element Switched Beam Array

multiple access techniques will be described in the Section "Wireless Multiplexing and Multiple Access Techniques, p. 87"). Smart antennas aim to overcome this limitation by allowing multiple nodes to transmit simultaneously, significantly increasing network throughput. There are two varieties of smart antenna — switched beam and adaptive array.

A switched beam antenna consists of an array of antenna elements each having a predefined beam pattern with a narrow main lobe and small side-lobes (Figure 3-17). Switching between beams allows one array element to be selected that provides the best gain in the direction of a target node, or the lowest gain towards an interfering source.

The simplest form of switched beam antenna is the pair of diversity receiver antennas often implemented in wireless LAN access points to reduce multipath effects in indoor environments. The receiver senses which of the two antennas is able to provide the highest signal strength and switches to that antenna.

Adaptive beams or beam-forming antennas consist of two or more antenna elements in an array and a so-called beam-forming algorithm, which assigns a specific gain and phase shift to the signal sent to or received from each antenna element. The result is an adjustable radiation

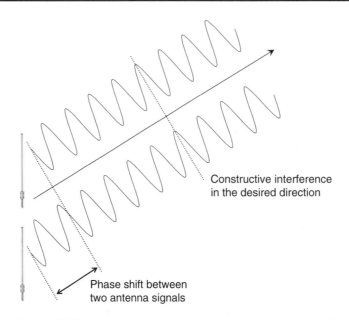

Constructive interference
in the desired direction

Phase shift between
two antenna signals

Figure 3-18: Phase Shift Between Two Antennas Resulting in a Directed Beam

pattern that can be used to steer the main lobe of the beam in the direction of the desired maximum gain (Figure 3-18).

As well as focussing its beam pattern towards a particular node, the adaptive beam antenna can also place a "null" or zero gain point in the direction of a source of interference. Because the gain and phase shift applied to individual array elements is under real-time software control (Figure 3-19), the antenna can dynamically adjust its beam pattern to compensate for multipath and other sources of interference and noise.

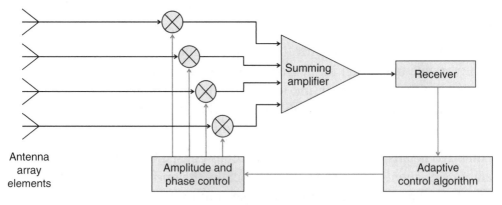

Antenna
array
elements

Summing
amplifier

Receiver

Amplitude and
phase control

Adaptive
control algorithm

Figure 3-19: Adaptive Beam Antenna

Like adaptive beam arrays, Multi-input Multi-output or MIMO radio, described in the Section "MIMO Radio, p. 124", also uses multiple antennas to increase network capacity. The key difference between these two techniques is that MIMO radio exploits multi-path propagation between a single transmitter and receiver, while adaptive beam arrays use multiple antennas to focus a single spatial channel. Some other differences are described in Table 3-6.

Table 3-6: Adaptive Beam Arrays and MIMO Radio Compared

	Adaptive beam array	*MIMO radio*
Objective	Focus the propagation pattern along a single desired spatial direction, to allow multiple access, reduce interference or increase range.	Exploit multi-path propagation to increase data capacity by multiplexing data streams over several spatial channels.
Antenna configuration	2 or more antenna elements at Tx and/or Rx. Tx and Rx configurations are independent.	Typically 2×2 or 4×4 (Tx × Rx). Tx and Rx are linked by the digital signal processing algorithm.
Spatial diversity	Single spatial channel focussed between transmitter and receiver.	Multiple spatial channels, exploiting multi-path propagation.
Data multiplexing	Single bit stream encoded to all transmit antennas.	Data stream multiplexed over spatial channels.
Signal processing	Simple phase and gain modification for each antenna.	Complex processing algorithm to decode signals over multiple spatial channels.
Application example	3rd generation WLAN access points — Section "Wireless LAN Arrays, p. 52".	PHY layer for the 802.11n standard — Section "MIMO and data rates to 600 Mbps (802.11n), p. 165".

Also under development is a new type of switched beam wireless LAN antenna called a plasma antenna, which uses a solid-state plasma (an ionised region in a silicon layer) as a reflector to focus and direct the emitted RF beam. A plasma antenna will be able to switch a medium gain (10–15 dBi) beam with approximately 10° beamwidth to 1 of 36 directions within a full 360° coverage, with a switching time that is less than the gap between transmitted frames.

Wireless PAN Devices

Wireless PAN Hardware Devices

Bluetooth Devices

In Part IV the wide range of PAN technologies will be described, from Bluetooth, which is most commonly identified with personal area networking, to ZigBee, an emerging technology primarily aimed at networking home and industrial control devices.

In fact with high powered (Class 1) Bluetooth radios, which can equal the range of Wi-Fi devices, the boundary between personal and local area networking is blurred and, within the limitations of achievable data rates, most of the WLAN devices described in the Section "Wireless LAN Devices p. 45" could equally well be built using Bluetooth technology.

The most common types of Bluetooth devices and some of their key features are summarised in Table 3-7 and shown in Figure 3-20.

Figure 3-20: A Variety of Bluetooth Devices (courtesy of Belkin Corporation, D-Link (Europe) Ltd., Linksys (a division of Cisco Systems Inc.) and Zoom Technologies, Inc.)

As developing wireless PAN technologies such as wireless USB mature, a comparable range of devices will be developed to support these networks. Novel capabilities inherent in these new technologies will also result in new device types offering new services, an example being the

Table 3-7: Bluetooth Devices and Features

Bluetooth device	*Key features*
Mobile phone	Interface with a Bluetooth hands-free headset. Connect to PDA or PC to transfer or back-up files. Exchange contact details (business cards), calendar entries, photos, etc. with other Bluetooth devices.
PDA	Connect to PC to transfer or back-up files. Connect to the Internet via a Bluetooth access point. Exchange contact details (business cards), calendar entries, photos, etc. with other Bluetooth devices.
Headset or headphones	Hands-free mobile telephony. Audio streaming from PC, TV, MP3 player or hi-fi system.
Audio transceiver	Audio streaming from a PC or hi-fi system to Bluetooth headphones.
Access point	Extend a LAN to include Bluetooth enabled devices. Internet connectivity for Bluetooth devices.
Bluetooth adapters	Bluetooth enable a range of devices, such as laptops or PDAs. As for WLAN NICS, they are available in a range of form factors, with USB dongles being the most popular. Serial adapter for plug-and-play connectivity to any serial RS-232 device.
Print adapter	Print files or photos from Bluetooth enabled device.
PC input devices	Wireless connectivity to a PC mouse or keyboard.
GPS receiver	Provide satellite navigation capabilities to Bluetooth-enabled devices loaded with required navigation software.
Dial-up modem	Provide wireless connectivity from a PC to a dial-up modem.

capability of the multi-band OFDM radio to spatially locate a wireless USB station, offering the potential for devices that rely on location based services.

ZigBee Devices
ZigBee is an emerging low data rate, very low power, wireless networking technology, described in the Section "ZigBee (IEEE 802.15.4), p. 273", that will initially focus on home automation but is likely to find a wide

61

range of applications, including a low cost replacement for Bluetooth in applications that do not require higher data rates.

The key features of a range of currently available and expected ZigBee devices is summarised in Table 3-8, and some of these devices are shown in Figure 3-21.

Wireless PAN antennas

In practice, since PAN operating range is generally under ten metres, Bluetooth and other PAN devices will typically use simple integrated omnidirectional antennas. However, for PANs such as Bluetooth that share the 2.4 GHz ISM radio band with 802.11b/g WLANs, the wide range of external WLAN antennas described above are also available to enable PAN devices to be operated with extended range.

Table 3-8: ZigBee Devices and Features

ZigBee device	Key features
PC input devices	Wireless connectivity to a PC mouse or keyboard.
Automation devices	Wireless control devices for home and industrial automation functions such as heating, lighting and security.
Wireless remote control	Replacing Ir remote for TV etc., and eliminating the line-of-sight and alignment restriction.
Sensor modem	Provides a wireless networking interface for a number of existing current loop sensors for home or industrial automation.
Ethernet gateway	A ZigBee network coordinator that enables command of ZigBee end devices or routers from an Ethernet network.

Wireless MAN Devices

While wireless LANs and PANs present a wide diversity of topologies and device types, to date wireless MAN devices have serviced only fixed point-to-point and point to multi-point topologies, requiring in essence only two device types, the base station and the client station.

However, following the ratification of the 802.16e standard (also designated 802.16-2005), broadband Internet access will soon be widely

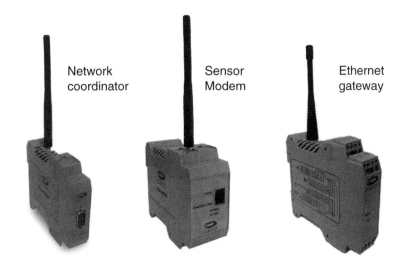

Network coordinator   Sensor Modem   Ethernet gateway

Figure 3-21: A Variety of ZigBee Devices (courtesy of Cirronet Inc.)

available to mobile devices and a range of new mobile wireless MAN devices is emerging, driving the convergence of mobile phones and PDAs.

Fixed Wireless MAN Devices

Wireless networking devices for fixed wireless MAN applications, essentially for last mile broadband Internet access, fall into two categories — base station equipment and customer premises equipment (CPE).

Some examples of base station equipment to support wireless MANs of differing scales are shown in Figure 3-22. The macro scale base station shown can potentially support thousands of subscribers in dense

Figure 3-22: Micro and Macro WMAN Base Station Equipment (courtesy of Aperto Networks Inc.)

Table 3-9: Wireless MAN Devices and Features

WMAN device	*Typical features*
Basic self installed indoor CPE	Basic WMAN connectivity to a customer PC or network. Multiple diversity or adaptive array antennas to improve non line-of-sight reception.
Outdoor CPE	External antenna and radio. Provides higher antenna gain and longer range.
Base station equipment	Modular and scalable construction. Macro and micro configurations for dense metropolitan or sparse rural installations. Flexible RF channel usage, from one channel over multiple antenna sectors to multiple channels per antenna sector.
Integrated network gateway	MAN interface with network gateway functions (Routing, NAT and firewall capabilities). Optionally with integrated wireless LAN access point. Integrated CPU to support additional WISP services such as VoIP telephony.

metropolitan area deployments, while the micro scale equipment is designed to support lower user numbers in sparse rural areas. Some of the types and key features of base station and CPE hardware are summarised in Table 3-9.

Figure 3-23 shows a variety of different wireless MAN CPE equipment.

Fixed Wireless MAN Antennas

Antennas for fixed wireless MAN applications similarly divide into base station and CPE. The general types of antennas summarised earlier in Table 3-5 for LANs are equally applicable to MAN installations, with appropriate housings for outdoor service and mountings designed for wind and ice loading. The factors that determine the choice of antenna are summarised in Table 3-10.

Depending on its elevation relative to the target area, a base station may comprise two sets of sector antennas as illustrated in Figure 3-24. A set of intermediate gain antennas, with higher vertical beamwidth, provide coverage over short-to-medium distances, with a second set of high-gain, low vertical beamwidth antennas providing coverage at longer range.

Figure 3-23: Fixed Wireless MAN CPE Equipment (courtesy of Aperto Networks Inc.)

Table 3-10: Factors Determining Wireless MAN Antenna Choice

Location	*Antenna type*	*Application*
CPE	Omnidirectional	Low gain requirement — for subscribers located close to the base station.
	Patch	Intermediate gain — mid range equipment should be applicable to most subscribers.
	Directional (Yagi or parabolic reflector)	High gain, high cost equipment to maximise data rate at the edge of the operating area.
Base station	Sector, intermediate gain	Wide area coverage close to the base station. Wider vertical beamwidth required to provide coverage close to the base station.
	Sector, high gain	Wide area coverage at a distance from the base station. High gain adds range with narrower vertical beamwidth.
	Directional	High gain antenna for point-to-point applications, such as backhaul, bridging between base stations, etc.

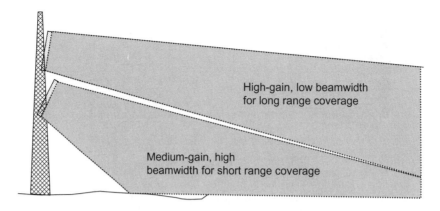

High-gain, low beamwidth
for long range coverage

Medium-gain, high
beamwidth for short range coverage

Figure 3-24: WMAN Base Station Sector Antenna Configuration

Mobile Wireless MAN Devices

The first implementation of mobile wireless MAN services and devices, delivering broadband Internet access to the user on the move, has been in the South Korean market, driven by the rapid development of the WiBro standard (a sub-set of 802.16e, described in the Section "TTA WiBro, p. 320"). Commercial uptake has also been speeded by the use of licensed spectrum, and the granting in 2005 of operating licences to three telecom companies.

The form factor for devices to deliver mobile Internet services reflects the need to combine telephony and PDA capabilities — a larger screen and QWERTY input to overcome the limitations experienced with WAP phones. Figure 3-25 shows two early WiBro phones developed by Samsung.

Summary of Part I

Chapters 2 and 3 have introduced the basic logical and physical architecture of wireless networks, the software and hardware elements that are the building blocks for the construction and operation of all wireless networks.

Although all networks require protocols and standards operating at all layers in the OSI model in order to operate, it is primarily the protocols and mechanisms of the Data Link and PHY layers which distinguish wired from wireless networking because of their specific design to

Figure 3-25: Wireless MAN Enabled Phones (courtesy of Samsung Electronics)

address the challenges of data transmission across the wireless medium. These protocols and mechanisms are explored in greater detail in Parts III to V, where the main wireless networking standards, such as 802.11 (WLAN), 802.15 (WPAN) and 802.16 (WMAN) are discussed.

Chapter 3 has also provided an overview of the different types of devices available for wireless networking at these three operating scales — personal, local and metropolitan. Device convergence is a common feature at the personal-local and local-metropolitan interfaces, and devices are increasingly appearing with multi-mode radios, enabling a single device to participate in a variety of different wireless networks.

WIRELESS COMMUNICATION

Introduction

The OSI network model illustrates how data and protocol messages from the application level cascade down through the logical layers and result in a series of data frames to be transmitted across the physical network medium.

In a wireless network that physical layer is provided by radio frequency (RF) or infrared (Ir) communications and in Part II the basics of these methods of wireless communication will be covered.

Starting with the RF spectrum, the regulation of spectrum use is briefly described and spread spectrum techniques are then introduced. This is a key technology that enables high data link reliability by making RF communications less susceptible to interference. Multiple access methods that enable many users to simultaneously use the same communication channel are then discussed. Signal coding and modulation is the step that encodes the data stream onto the RF carrier or pulse train, and a range of coding and modulation techniques applied in wireless networking will be covered, from the simplest to some of the most complex.

The various elements that impact on RF signal propagation will be described, enabling a calculation of the link budget — the balance of power available to overcome system and propagation losses to bring the transmitted signal to the receiver at a sufficient power level for reliable, low error rate reception. The link budget calculation is an essential part of the toolkit of the wireless network designer in defining the basic power requirements for a given network installation. This will usually be supplemented by practical techniques such as site surveys, which will be covered in Parts III and V.

Ultra wideband (UWB) radio has emerged as a key technology for short range wireless network applications, and some of the varieties of UWB radio that are applied in wireless USB, wireless FireWire and ZigBee will be described.

Chapter 5 gives a similar overview of aspects of infrared communications, as used in IrDA connections, covering the infrared spectrum, Ir propagation and reception, and Ir link calculation.

CHAPTER 4

Radio Communication Basics

The RF Spectrum

The radio frequency, or RF, communication at the heart of most wireless networking operates on the same basic principles as everyday radio and TV signals. The RF section of the electromagnetic spectrum lies between the frequencies of 9 kHz and 300 GHz (Table 4-1), and different bands in the spectrum are used to deliver different services.

Recalling that the wavelength and frequency of electromagnetic radiation are related via the speed of light, so that wavelength (λ) = speed of light (c) / frequency (f), or wavelength in metres = 300 / frequency in MHz.

Table 4-1: Subdivision of the Radio Frequency Spectrum

Transmission type	Frequency	Wavelength
Very low frequency (VLF)	9–30 kHz	33–10 km
Low frequency (LF)	30–300 kHz	10–1 km
Medium frequency (MF)	300–3000 kHz	1000–100 m
High frequency (HF)	3–30 MHz	100–10 m
Very high frequency (VHF)	30–300 MHz	10–1 m
Ultra high frequency (UHF)	300–3000 MHz	1000–100 mm
Super high frequency (SHF)	3–30 GHz	100–10 mm
Extremely high frequency (EHF)	30–300 GHz	10–1 mm

71

Beyond the extremely high frequency (EHF) limit of the RF spectrum lies the infrared region, with wavelengths in the tens of micrometre range and frequencies in the region of 30 THz (30,000 GHz).

Virtually every Hz of the RF spectrum is allocated for one use or another (Figure 4-1), ranging from radio astronomy to forestry conservation, and some RF bands have been designated for unlicensed transmissions.

The RF bands which are used for most wireless networking are the unlicensed ISM or Instrument, Scientific and Medical bands, of which the three most important lie at 915 MHz (868 MHz in Europe), 2.4 GHz and

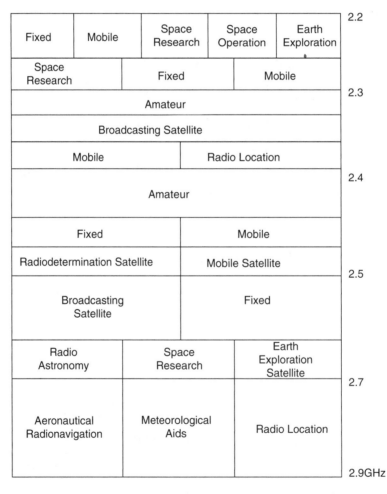

Figure 4-1: FCC Spectrum Allocation Around the 2.4GHz ISM Band

Table 4-2: Radio Frequency Bands in Use for Wireless Networking

RF band	*Wireless networking specification*
915/868 MHz ISM	ZigBee
2.4 GHz ISM	IEEE 802.11b, g, Bluetooth, ZigBee
5.8 GHz	IEEE 802.11a

5.8 GHz (Table 4-2). As well as these narrow band applications, new networking standards such as ZigBee (Section "ZigBee (IEEE 802.15.4), p. 273)" will make use of the FCC spectrum allocation for ultra wideband radio (UWB — see Section "Ultra Wideband Radio p. 119") that permits very low power transmission across a broad spectrum from 3.1 to 10.6 GHz.

Radio Frequency Spectrum Regulation

The use of the radio frequency spectrum, in terms of the frequency bands that can be used for different licensed and unlicensed services, and the allowable transmission power levels for different signal formats, are controlled by regulatory authorities in individual countries or regions (Table 4-3).

Although there is an increasing trend towards harmonisation of spectrum regulation across countries and regions, driven by the International Telecommunications Union's World Radio Communication Conference, there are significant differences in spectrum allocation and other conditions such as allowable transmitter power levels which have an impact on wireless networking hardware design and interoperability.

As an example, in the 5.8 GHz ISM band used for IEEE 802.11a networks, the FCC in the USA allows a maximum transmitted power of

Table 4-3: Radio Frequency Spectrum Regulatory Bodies

Country/Region	*Regulatory body*	
USA	FCC	Federal Communications Commission
Canada	IC	Industry Canada
Europe	ETSI	European Telecommunications Standards Institute
Japan	ARIB	Association of Radio Industries and Businesses

Table 4-4: 2.4 GHz ISM Band Regulatory Differences by Region

Regulator	*2.4 GHz ISM specifications*
FCC (USA)	1 W maximum transmitted power 2.402–2.472 GHz, 11 × 22 MHz channels
ETSI (Europe)	100 mW maximum EIRP 2.402–2.483 GHz, 13 × 22 MHz channels
ARIB (Japan)	100 mW maximum EIRP 2.402–2.497 GHz, 14 × 22 MHz channels

1 W, while in Europe the ETSI permits a maximum EIRP (equivalent isotropic radiated power) of just 100 mW EIRP or 10 mW/MHz of bandwidth, with variations in other countries. Table 4.4 shows a range of other regulatory differences that apply to the 2.4 GHz ISM band used for IEEE 802.b/g networks.

The pace of regulatory change also differs from region to region. For example, the FCC developed regulations governing ultra wideband radio in 2002, while in Europe ETSI Task Group 31 was still working on similar regulations in 2006.

Although the regulatory bodies impose conditions on the unlicensed use of parts of the RF spectrum, unlike their role in the licensed spectrum, these bodies take no responsibility for or interest in any interference between services that might result from that unlicensed use. In licensed parts of the RF spectrum, the FCC and similar bodies have a role to play in resolving interference problems, but this is not the case in unlicensed bands. Unlicensed means in effect that the band is free for all, and it is up to users to resolve any interference problems. This situation leads some observers to predict that the 2.4 GHz ISM band will eventually become an unusable junk band, overcrowded with cordless phone, Bluetooth, 802.11 and a cacophony of other transmissions. This impending "tragedy of the commons" may be prevented by the development of spectrum agile radios, described in Chapter 14.

Further information on current spectrum regulation and future developments, including the further development of regulations on ultra

Table 4-5: Web Sites of Spectrum Regulators

Regulator	*Country/Region*	*URL*
FCC	USA	www.fcc.gov
Industry Canada	Canada	www.ic.gc.ca
ETSI	Europe	www.etsi.org
ARIB	Japan	www.arib.or.jp/english

wideband radio outside the USA, can be found from the regulators web sites at the URLs shown in Table 4-5.

Radio Transmission as a Network Medium

Compared to traditional twisted-pair cabling, using RF transmission as a physical network medium poses a number of challenges, as outlined in Table 4-6. Security has been a significant concern since RF transmissions are far more open to interception than those confined to a cable. Security issues will be covered in detail in Chapters 8 and 11.

Data link reliability, bit transmission errors resulting from interference and other signal propagation problems, are probably the second most significant challenge in wireless networks, and one technology that resulted in a quantum leap in addressing this problem (spread spectrum transmission) is the subject of the next section.

Controlling access to the data transmission medium by multiple client devices or stations is also a different type of challenge for a wireless medium, where, unlike a wired network, it is not possible to both transmit and receive at the same time. Two key situations that have the potential to

Table 4-6: The Radio Frequency Networking Challenge

Challenges	*Considerations and solutions*
Link reliability	Signal propagation, interference, equipment siting, link budget.
Media access	Sensing other users (hidden station and exposed station problems), Quality of service requirements.
Security	Wired equivalent privacy (WEP), Wi-Fi Protected Access (WPA), 802.11i, directional antennas.

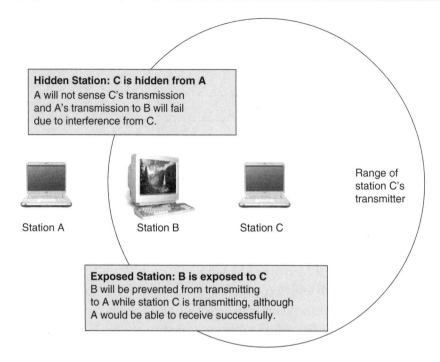

Figure 4-2: Hidden and Exposed Station Challenges for Wireless Media Access Control

degrade network performance are the so-called hidden station and exposed station problems.

The hidden station problem occurs when two stations A and C are both trying to transmit to an intermediate station B, where A and C are out of range and therefore one cannot sense that the other is also transmitting (see Figure 4-2). The exposed station problem occurs when a transmitting station C, prevents a nearby station B from transmitting although B's intended receiving station A is out of range of station C's transmission.

The later sections of this chapter look at digital modulation techniques and the factors affecting RF propagation and reception, as well as the practical implications of these factors in actual wireless network installations.

Spread Spectrum Transmission

Spread spectrum is a radio frequency transmission technique initially proposed for military applications in World War II with the intention of making wireless transmissions safe from interception and jamming. These techniques started to move into the commercial arena in the early

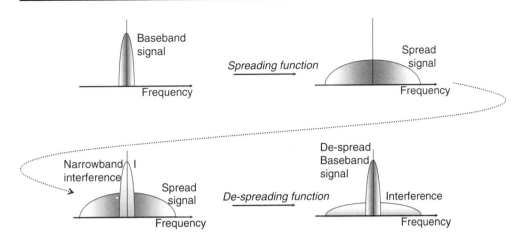

Figure 4-3: A Simple Explanation of Spread Spectrum

1980s. Compared to the more familiar amplitude or frequency modulated radio transmissions, spread spectrum has the major advantage of reducing or eliminating interference with narrowband transmissions in the same frequency band, thereby significantly improving the reliability of RF data links.

Unlike simple amplitude or frequency modulated radio, a spread spectrum signal is transmitted using a much greater bandwidth than the simple bandwidth of the information being transmitted. Narrow band interference (the signal I in Figure 4-3) is rejected when the received signal is "de-spread". The transmitted signal also has noise-like properties and this characteristic makes the signal harder to eavesdrop on.

Types of Spread Spectrum Transmission

The key to spread spectrum techniques is some function, independent of the data being transmitted, that is used to spread the information signal over a wide transmitted bandwidth. This process results in a transmitted signal bandwidth which is typically 20 to several 100 times the information bandwidth in commercial applications, or 1000 to 1 million times in military systems.

Several different methods of spread spectrum transmission have been developed, which differ in the way the spreading function is applied to the information signal. Two methods, direct sequence spread spectrum and frequency hopping spread spectrum, are most widely applied in wireless networking.

In Direct Sequence Spread Spectrum (DSSS) (Figure 4-4), the spreading function is a code word, called a chipping code, that is XOR'd with the input bit stream to generate a higher rate "chip stream" that is then used to modulate the RF carrier.

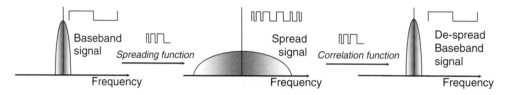

Figure 4-4: A Simple Explanation of DSSS

In Frequency Hopping Spread Spectrum (FHSS) (Figure 4-5), the input data stream is used directly to modulate the RF carrier while the spreading function controls the specific frequency slot of the carrier within a range of available slots spread across the width of the transmission band.

Time Hopping Spread Spectrum (THSS) (Figure 4-6), is a third technique in which the input data stream is used directly to modulate the RF carrier which is transmitted in pulses with the spreading function controlling the

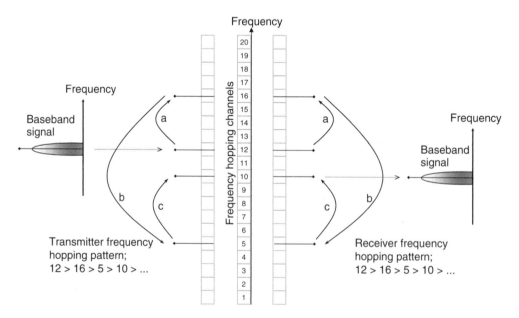

Figure 4-5: A Simple Explanation of FHSS

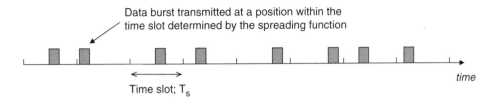

Figure 4-6: A Simple Explanation of THSS

timing of each data pulse. For example, impulse radio uses pulses that are so short, typically in the region of 1 nanosecond (nS), that the spectrum of the signal is very wide and meets the definition of an ultra wideband (UWB) system. The spectrum is effectively spread as a result of the narrowness of transmitted pulses, but time-hopping, with each user or node being assigned a unique hopping pattern, is a simple technique for impulse radio to allow multiple user access (see Section "Time Hopping PPM UWB (Impulse Radio), p. 121").

Two other less common techniques are Pulsed FM systems and Hybrid systems (Figure 4-7). In Pulsed FM systems, the input data stream is used directly to modulate the RF carrier, which is transmitted in frequency

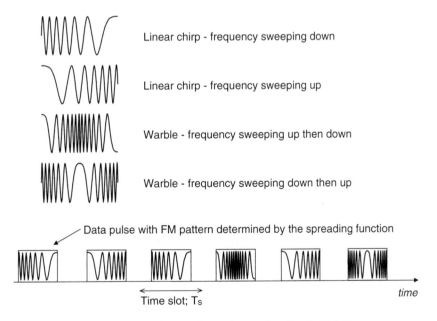

Figure 4-7: A Simple Explanation of Pulsed FM Systems

modulated pulses. The spreading function controls the pattern of frequency modulation, which could for example be a linear "chirp" with frequency sweeping up or down.

Hybrid systems also use combinations of spread spectrum techniques and are designed to take advantage of specific characteristics of the individual systems. For example, FHSS and THSS methods are combined to give the hybrid frequency division – time division multiple access (FDMA/TDMA) technique (see the Section "Wireless Multiplexing and Multiple Access Techniques, p. 87").

Of these alternative spread spectrum techniques, DSSS and FHSS are specified in the IEEE 802.11 wireless LAN standards, although DSSS is most commonly used in commercial 802.11 equipment. FHSS is used by Bluetooth, and FHSS and chirp spread spectrum are optional techniques for the IEEE 802.15.4a (ZigBee) specification.

Chipping, Spreading and Correlating

The spreading function used in DSSS is a digital code, known as a chipping code or pseudo-noise (PN) code, which is chosen to have specific mathematical properties. One such property is that, to a casual listener on the broadcast band, the signal is similar to random noise, hence the "pseudo-noise" label.

Under the IEEE 802.11b standard, the specified PN code for 1 Mbps and 2 Mbps data rates is the 11-bit Barker code. Barker codes are binary sequences that have low auto-correlation, which means that the sequence does not correlate with a time-shifted version of itself. Barker codes of length 2 to 13 are shown in Table 4-7.

Figure 4-8 illustrates the direct sequence encoding of a data stream using this code. To distinguish a data bit from a code bit, each symbol (each 1 or 0) in the coded sequence is known as a chip rather than a bit.

This process results in a chip stream with a wider bandwidth than the original input data stream. For example, a 2 Mbps input data rate is encoded into a 22 Mcps (Mega chips per second) sequence, the factor of 11 coming about since the encoded sequence has 11 chips for each data bit. The resulting bandwidth of the transmitted RF signal will depend on the technique used to modulate the encoded data stream onto the RF

Table 4-7: Barker Codes of Length 2 to 13

Length	Code
2	10 and 11
3	110
4	1011 and 1000
5	11101
7	1110010
11	11100010010
13	1111100111001

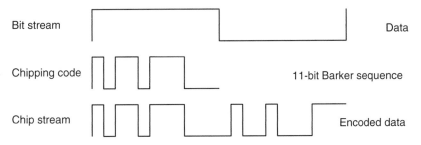

Figure 4-8: DSSS Pseudo-noise Encoding

carrier. As described in the Section "Digital Modulation Technique, p. 95", a simple modulation technique like binary phase-shift keying (BPSK) results in a modulated carrier with a bandwidth equal to twice the input bit rate (or in this case, the chip rate).

When the encoded signal is received, a code generator in the receiver recreates the same PN code and a correlator uses this to decode the original information signal in a process known as correlating or de-spreading. Since the correlator only extracts signals encoded with the same PN code, the receiver is unaffected by interference from narrow band signals in the same RF band, even if these signals have a higher power density (in watts/Hz) than the desired signal.

Chipping Codes

One of the desirable mathematical properties of PN codes is that it enables the receiver's PN code generator to very rapidly synchronise with the PN code in the received signal. This synchronisation is the first step in the de-spreading process. Fast synchronisation requires that the position of the code word can be quickly identified in a received signal, and this is achieved as a result of the low auto-correlation property of the Barker codes. Another benefit of low auto-correlation is that the receiver will reject signals that are delayed by more than one chip period. This helps to make the data link robust against multipath interference, which will be discussed in the Section "RF Signal Propagation and Losses, p. 112".

A second key property of chipping codes that is important in applications where interference between multiple transmitters must be avoided, for example in mobile telephony, is low cross-correlation. This property reduces the chance that a correlator using one PN code will experience interference from a signal using a different code (i.e. that it will incorrectly decode a noisy signal that was encoded using a different chipping code). Ideally codes in use in this type of multiple access application should have zero cross-correlation, a property of the orthogonal codes used in CDMA (Section "Code Division Multiple Access, p. 94").

Code orthogonality for multiple access control is not required for wireless networking applications, such as IEEE 802.11 networks, as these standards use alternative methods to avoid conflict between overlapping transmitted signals from multiple users, which are described in the Section "Wireless Multiplexing and Multiple Access Techniques, p. 87".

Complementary Code Keying

An alternative to using a single chipping code to spread every bit in the input data stream is to use a set of spreading codes and to select one code from the set depending on the values of a group of input data bits. This scheme is known as complementary code keying (CCK).

CCK was proposed to the IEEE by Lucent Technologies and Harris Semiconductor (now part of Intersil Corp.) in 1998, as a means to raise the IEEE 802.11b data rate to 11 Mbps. Instead of using the Barker code, they proposed to use a set of codes called Complementary Sequences, based on the Walsh/Hadamard transforms (see the Section "Code Division Multiple Access, p. 94").

Using CCK, a chipping code word is chosen from a set of 64 unique codes depending on the value of each 6-bit segment of the input data stream. The encoded data sequence comprises a series of code words, and this chip sequence is modulated onto the RF carrier using one of a variety of modulation techniques that will be described in the Section "Digital Modulation Techniques, p. 93".

The main advantage of CCK modulation is spectral efficiency, since each transmitted code word represents 6 input data bits instead of the single bit represented by the Barker code. CCK can achieve 11 Mbps using the same 22 MHz bandwidth used to transmit 1 Mbps with the Barker code. However, the price of this high data rate is complexity. A receiver using the Barker code requires just one correlator to pick out the chipping code, while a CCK system needs 64 correlators, one on the lookout for each of the complementary codes.

Direct Sequence Spread Spectrum in the 2.4 GHz ISM Band

As noted above, in DSSS the data signal is combined with a code word, the chipping code, and the combined signal is used to modulate the RF carrier, resulting in a transmitted signal spread over a wide bandwidth. For example, in the 2.4 GHz ISM band, a spread bandwidth of 22 MHz is specified for IEEE 802.11 networks, as shown in Figure 4-9.

The 2.4 GHz ISM band has a total allowed width of 83.5 MHz and is divided into a number of channels (11 in the USA, 13 in Europe, 14 in Japan), with 5 MHz steps between channels. To fit 11 or more 22 MHz

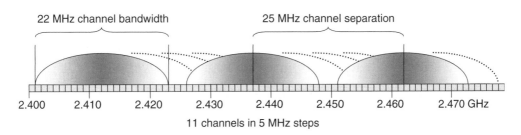

Figure 4-9: 802.11 DSSS Channels

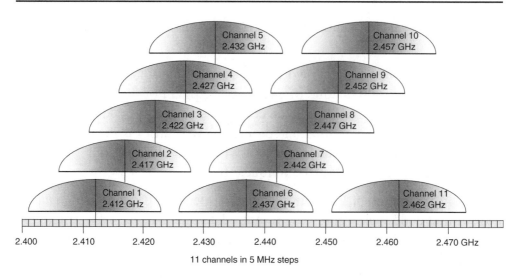

Figure 4-10: DSSS Channels in the 2.4 GHz ISM Band (US)

wide channels into an 83 MHz wide band results in considerable overlap between the channels (as shown in Figure 4-10), resulting in the potential for interference between signals in adjacent channels. The 3 non-overlapping channels allow 3 DSSS networks to operate in the same physical area without interference.

The use of DHSS in 802.11 wireless networks is described further in Chapter 6.

Frequency Hopping Spread Spectrum in the 2.4 GHz ISM Band

In frequency hopping spread spectrum transmission (FHSS) the data is modulated directly onto a single carrier frequency, but that carrier frequency hops across a number of channels within the RF band using a pseudo-random hopping pattern. In the 2.4 GHz ISM band for example, a maximum channel width of 1 MHz is specified for FHSS systems, and 79 such channels are available. A transmitter switches between these channels many times a second, moving on to the next channel in its sequence after a predetermined time, known as the "dwell time".

The IEEE 802.11b standard specifies that the hop must be to a new channel a minimum of 6 MHz from the previous channel, and that hops must occur at least 2.5 time per second (Figure 4-11). The spectrum regulators specify the allowable limits for transmission parameters such as

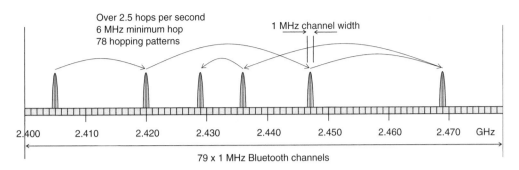

Over 2.5 hops per second
6 MHz minimum hop
78 hopping patterns

1 MHz channel width

2.400   2.410   2.420   2.430   2.440   2.450   2.460   2.470   GHz

79 x 1 MHz Bluetooth channels

Figure 4-11: FHSS Channels Within the 2.4Ghz ISM Band

the maximum dwell time and individual standards, like IEEE 802.11, have to work within these boundary conditions.

In the receiver, a PN generator recreates the same hopping pattern. This allows the receiver to make the same channel-to-channel hops as the transmitter, so that the data signal can be decoded.

Because the probability of two networks selecting the same channel at the same time is very low, many more FHSS networks can overlap physically without interference than is the case for DSSS networks.

Frequency hopping spread spectrum in the 2.4 GHz ISM band is specified alongside DSSS as an option in the IEEE 802.11b standard and is also used in Bluetooth networks.

All 79 available channels are normally used in Bluetooth, although an alternative hopping sequence that uses 23 channels (2.454 to 2.476 GHz), is available for use in France where special regulatory conditions apply. Frequency hops occur after each data packet, which will be a multiple of 1-, 3- or 5-times the time slot duration of 625 microseconds (320 to 1600 hops/second). The frequency hopping pattern is determined by the unique 48-bit device ID of the master device in each Bluetooth piconet, and synchronisation to the hopping pattern is part of the process of device discovery when a new device joins the piconet. The use of FHSS in Bluetooth is described further in Chapter 10.

Time Hopping Spread Spectrum
In a time hopping spread spectrum system, time is divided into frames, with each frame divided into a number of transmission slots. Within each

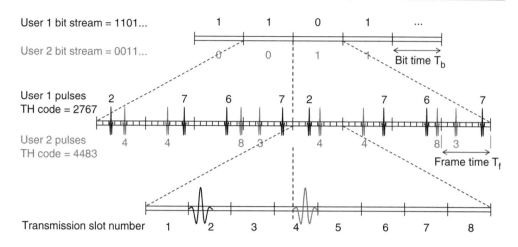

Figure 4-12: Time Hopping Spread Spectrum

frame, data is transmitted only during one time slot and the specific time slot used during each frame is determined using a PN code. Figure 4-12 shows a THSS system with two clients using different hopping codes.

Impulse radio is an ultra-wide band transmission technique that is a candidate for the IEEE 802.15.4a (ZigBee) physical layer specification. This is a time hopping spread spectrum technique where a very short pulse is transmitted in each transmission time slot. Information is encoded via pulse position or pulse amplitude modulation (PPM, PAM). The spreading effect of time hopping, together with the short pulse duration, results in a transmitted signal spread across an ultra-wide bandwidth.

Table 4-8: Benefits of Common Spread Spectrum Techniques

Frequency Hopping	*Direct sequence*
Simple to design and manufacture	Higher data speeds
Cheaper to implement	Increased range
Higher density of overlapping networks	Throughput is interference-tolerant up to a threshold level
Gradual degradation of throughput with interference	

Spread Spectrum in Wireless Networks — Pros and Cons

The advantages of spread spectrum techniques, such as resistance to interference and eavesdropping and the ability to accommodate multiple users in the same frequency band, make this an ideal technology for wireless network applications (Table 4-8). Although the good interference performance is achieved at the cost of relatively inefficient bandwidth usage, the available radio spectrum, such as the 2.4 GHz ISM band, still permits data rates of up to 11 Mbps using these techniques.

Since speed and range are important factors in wireless networking applications, DSSS is the more widely used of the two techniques although, because of its simpler and cheaper implementation, FHSS is used for lower rate, shorter-range systems like Bluetooth and the now largely defunct HomeRF.

Wireless Multiplexing and Multiple Access Techniques

Introduction

Multiplexing techniques aim to increase transmission efficiency by transmitting multiple signals or data streams on a single medium. The resulting increased capacity can be used either to deliver a higher data rate to a single user, or to allow multiple users to access the medium simultaneously without interference.

User access to the bandwidth can be separated by a numbers of means: in time (TDMA), in frequency (FDMA or OFDMA), in space (SDMA) or by assigning users unique codes (CDMA). These methods will be described in turn in the following sections.

Time Division Multiple Access

Time division multiple access (TDMA) allows multiple users to access a single channel without interference by allocating specific time slots to each user. As shown in Figure 4-13, the time axis is divided into time slots that are assigned to users according to a slot allocation algorithm.

A simple form of TDMA is time division duplex (TDD), where alternate transmit periods are used for uplink and downlink in a duplex communication system. TDD is used in cordless phone systems to accommodate two-way communication in a single frequency band.

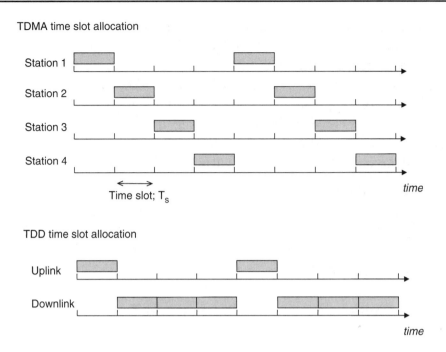

Figure 4-13: Time Division Multiple Access (TDMA) and Duplexing (TDD)

TDMA is used in Bluetooth piconets (see Chapter 10). The master device provides the system clock that determines the timing of slots and, within each time slot, the master first polls slave devices to see which devices need to transmit and then allocates transmission time slots to devices that are ready to transmit.

Frequency Division Multiple Access

In contrast to TDMA, frequency division multiple access (FDMA) provides each user with a continuous channel that is restricted to a fraction of the total available bandwidth. This is done by dividing the available bandwidth into a number of channels that are then allocated to individual users as shown in Figure 4-14.

Frequency division duplex (FDD) is simple form of FDMA in which the available bandwidth is divided into two channels to provide continuous duplex communication. Cellular phone systems such as GSM (2G) and UMTS (3G) use FDD to provide separate uplink and downlink channels, while 1G cellular phone systems used FDMA to allocate bandwidth to multiple callers.

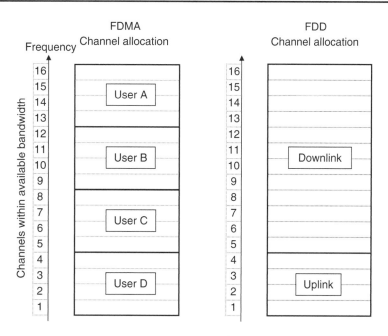

Figure 4-14: Frequency Division Multiple Access (FDMA) and Duplexing (FDD)

In practice, FDMA is often used in combination with TDMA or CDMA to increase capacity on a single channel in an FDMA system. As shown in Figure 4-15, FDMA/TDMA divides the available bandwidth into channels and then divides each channel into time slots that are allocated to individual users.

FDMA/TDMA is used by GSM cellular phones, with eight time slots available in each 200 kHz radio channel.

Orthogonal Frequency Division Multiplexing

Orthogonal frequency division multiplexing (OFDM) is a variant of frequency division multiplexing (FDM), in which a number of discrete subcarrier frequencies are transmitted within a band with frequencies chosen to ensure minimum interference between adjacent subcarriers.

This is achieved by controlling the spectral width of the individual subcarriers (also called tones) so that the frequencies of subcarriers coincide with minima in the spectra of adjacent subcarriers, as shown in Figure 4-16.

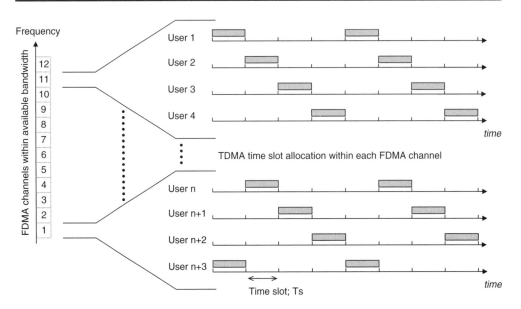

Figure 4-15: FDMA/TDMA Multiple Access System as Used in GSM Cellular Phones

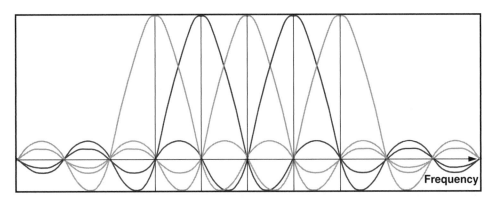

Figure 4-16: Orthogonality of OFDM Subcarriers in the Frequency Domain

In the time domain, the orthogonality of OFDM tones means that the number of subcarrier cycles within the symbol transmission period is an integer, as illustrated in Figure 4-17. This condition can be expressed as:

$$T_s = n_i \, / \, v_i \quad \text{or} \quad v_i = n_i \, / \, T_s$$

where T_s is the symbol transmission period and v_i is the frequency of the ith subcarrier. The subcarriers are therefore evenly spaced in frequency, with separation equal to the reciprocal of the symbol period.

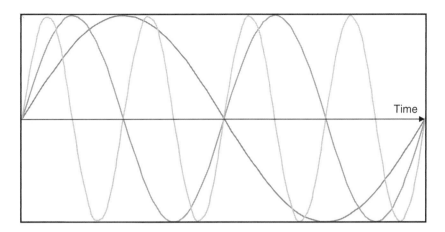

Figure 4-17: Orthogonality of OFDM Subcarriers in the Time Domain

There are a number of ways in which the multiple subcarriers of OFDM can be used:

- OFDM can be used as a multiple access technique (OFDMA), by assigning single subcarriers or groups of subcarriers to individual users according to their bandwidth needs.

- A serial bit stream can be turned into a number of parallel bit streams each one of which is encoded onto a separate subcarrier. All available subcarriers are used by a single user to achieve a high data throughput.

- A bit stream can be spread using a chipping code and then each chip can be transmitted in parallel on a separate subcarrier. Since the codes can allow multiple user access, this system is known as Multi-Carrier CDMA (MC-CDMA). MC-CMDA is under consideration by the WIGWAM project as one of the building blocks of the 1 Gbps wireless LAN (see the Section "Gigabit Wireless LANs, p. 350").

A significant advantage of OFDM is that, since the symbol rate is much lower when spread across multiple carriers than it would be if the same total symbol rate were transmitted on a single carrier, the wireless link is much less susceptible to inter-symbol interference (ISI). ISI occurs when, as a result of multi-path propagation, two symbols transmitted at different times arrive together at the receiving antenna after traversing different propagation paths (Figure 4-18). Although OFDM is inherently less

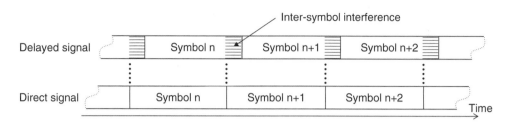

Figure 4-18: Inter Symbol Interference (ISI)

susceptible to ISI, most OFDM systems also introduce a guard interval between each symbol to further reduce ISI.

OFDM radios also use a number of subcarriers, called pilot tones, to gather information on channel quality to aid demodulation decisions. These subcarriers are modulated with known training data at the start of each transmitted data packet. Decoding this known data enables the receiver to determine and adaptively correct for the frequency offset and phase noise between the reference oscillators in the transmitter and the receiver and for fading during propagation.

Figure 4-19 shows a schematic block diagram of a simple OFDM transmitter and receiver. From the left, the input bit stream at a rate of R bps passes through a series to parallel converter and is split into N bit stream of rate R/N bps. Each of these bit streams drives one modulator, which maps each bit or symbol onto a point in the modulation constellation being used (Section "Digital Modulation Technique, p. 95"). The N resulting amplitude and phase points drive the inputs of an Inverse Fast Fourier Transform (IFTT), the output of which is the sum of the subcarriers, each modulated according to the individual input bit streams.

At the receiver, after removing any guard interval, a Fast Fourier Transform (FFT) determines the amplitude and phase of each subcarrier in the received signal. The amplitude and phase are adjusted using information gathered from the pilot tones. A demodulation decision is made by mapping this amplitude and phase onto the modulation constellation and the corresponding input bit or bits are generated. The resulting N parallel R/N bps bit streams are then combined in a parallel to series converter to give the original R bps bit stream.

The IEEE 802.11a/g standards uses OFDM in the unlicensed 2.4 and 5 GHz ISM bands respectively to provide data rates up to 54 Mbps.

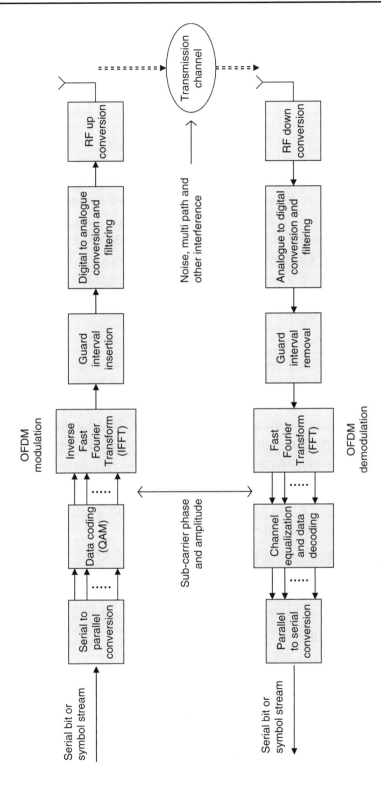

Figure 4-19: Schematic Block Diagram of an OFDM Transmitter and Receiver

The system uses 52 subcarriers of which 48 are used to carry data and are modulated using binary or quadrature phase shift keying (BPSK/QPSK), 16-quadrature amplitude modulation (QAM) or 64-QAM. The remaining four subcarriers are used as pilot tones.

Space Division Multiple Access

Space division multiple access (SDMA) is a technique which aims to multiply the data throughput of a wireless network by using spatial position as an additional parameter to control user access to the transmission medium. As a simple example, if a base station is equipped with sector antennas with a 30° horizontal beamwidth, it can separate users into twelve spatial segments or channels depending on their location around the base station. (Figure 3-17 shows the beam pattern for a base station with 6 such elements.) This arrangement would enable the network to achieve a potential twelve-fold increase in data capacity compared with a base station using a single isotropic antenna.

As well as simple sector antennas, smart antenna systems are being developed which combine an array of antennas with digital signal processing capabilities in order to achieve spatial control of transmission and reception. Smart antenna systems can adapt their directional characteristics in response to the signal environment and system demands, and can provide the basis for SDMA.

Generally a second multiple access technique, such as TDMA or CDMA, is also used in combination with SDMA in order to allow multiple user access within a single spatial segment.

Space division multiplexing (SDM), as opposed to SDMA, is based on the use of multiple propagation paths to simultaneously transmit multiple data channels using the same RF spectrum. This is the basis of MIMO radio (see the Section "MIMO Radio, p. 124") which is specified in the IEEE 802.11n standard.

Code Division Multiple Access

CDMA is closely related to DSSS, where a pseudo-noise code is used to spread a data signal over a wide bandwidth in order to increase its immunity to interference. As noted above, if two or more transmitters use different, orthogonal pseudo-noise (PN) codes in DS spread spectrum,

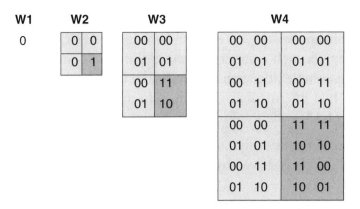

Figure 4-20: Construction of the Walsh Codes

they can operate on the same frequency band and in the same physical area without interfering. This is because a correlator using one PN code will not detect a signal encoded using another orthogonal code, since orthogonal codes by definition do not correlate with each other.

Examples of an orthogonal code set are the Walsh codes, which can be easily generated from the procedure called the Hadamard transform. With each step to the right in Figure 4-20, the three light matrices are the same as the full matrix to the left, while the darker shaded matrix is the inverse of the one to the left. The Walsh codes can be read off as the lines in each matrix, so the Walsh codes of length 4 are; 0000, 0101, 0011 and 0110.

The property of orthogonality is the basis of CDMA and is used in 3G mobile telephony to ensure that many users, each assigned a unique orthogonal access code, can transmit and receive without interference within a single network cell.

Digital Modulation Technique

Introduction

Modulation is the step in the digital signal processing sequence that transforms and encodes the data stream onto the transmitted RF or infrared signal. The spectrum spreading and multiple access techniques will result in a bit-stream transformed into a chip-stream which must now be modulated onto either a single or multiple carrier frequencies, or used to modulate the position or shape of a transmitted RF or Ir pulse.

A wide variety of modulation techniques are used in wireless networking. These range from the simple return to zero inverted (RZI) used in IrDA at low data rates, through a variety of phase shift and code keying methods of increasing complexity, such as BPSK and CCK, used for example in IEEE 802.11b at intermediate data rates, to more complex methods, such as the HHH (1,13) code used in IrDA at high data rates.

The selection of the best digital modulation technique for a specific application is driven by a number of criteria, the most important being:

- Spectral efficiency — achieving the desired data rate within the available spectral bandwidth (see Table 4.9).

- Bit error rate (*BER*) performance — achieving the required error rate given the specific factors causing performance degradation in the particular application (interference, multipath fading, etc.).

- Power efficiency — particularly important in mobile applications where battery life is an important user acceptance factor.

- Modulation schemes with higher spectral efficiency (in terms of data bits per Hz of bandwidth) require higher signal strength for error-free detection.

- Implementation complexity — which translates directly into the cost of hardware to apply a particular technique. Some aspects of modulation complexity can be implemented in software, which has less impact on end-user costs.

Table 4-9: Spectral Efficiency of Typical Modulation Techniques

Modulation technique	*Spectral efficiency (Bits/Hz)*
BPSK	0.5
QPSK	1.0
16-QAM	2.0
128-QAM	3.5
256-QAM	4.0

Simple Modulation Techniques

On/Off keying (OOK) is perhaps the simplest modulation technique, where the carrier is turned off during a 0-bit and turned on during a 1-bit. OOK is a special case of amplitude shift keying (ASK) in which two amplitude levels represent 0- and 1-bits. The magnitude of the amplitude shift between these two levels is called the modulation index.

Return to zero inverted (RZI) is the modulation technique used in IrDA for data rates up to 1.152 Mbps. It is a derivation of the non-return to zero (NRZ) modulation used in UART data transmission (Figure 4-21), in which a 1-bit is represented by a high state, a 0-bit by a low state, and the transition from high state to low state only occurs when a 1-bit is followed by a 0-bit.

In contrast, a return to zero (RZ) transmission has a low-high-low pulse during the bit time for each 1-bit, while the RZI scheme inverts this to give a pulse for each zero bit or symbol.

When the RZI modulated signal is received, the bit stream is recovered by triggering a high to low transition for each received pulse, as shown in Figure 4-22. The low state of the decoded signal returns to a high state and a high state remains high at the end of each bit period, unless another pulse is received.

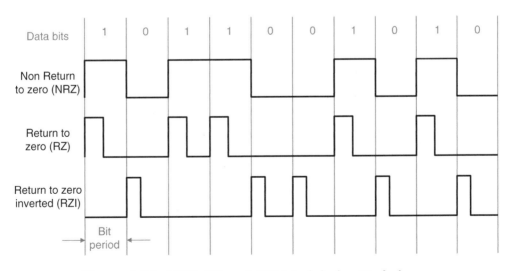

Figure 4-21: NRZ, RZ and RZI Modulation Techniques

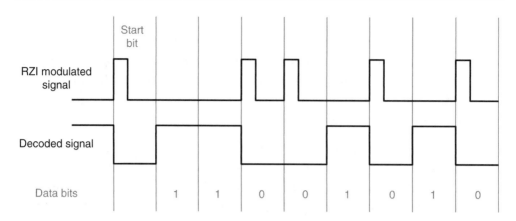

Figure 4-22: RZI Bit Stream Decoding

The advantage of the RZI scheme for IrDA is that it allows the transmitting LED to be off for most of the bit time, in order to conserve battery power.

Phase Shift Keying
Phase shift keying is a modulation technique in which the phase of the carrier is determined by the input bit or chip stream. There are several types of phase shift key (PSK) modulation including binary (BPSK) and quadrature (QPSK).

Binary Phase Shift Keying
BPSK is the simplest technique in this class, with the carrier phase taking one of two states, as shown in Table 4-10. A 0 input symbol (whether it is a bit or a chip) corresponds to a zero phase carrier while a 1-symbol corresponds to a 180° phase shifted carrier, resulting in the output waveform shown in Figure 4-23.

Table 4-10: Binary Phase Shift Keying

Symbol	Carrier phase
0	0 degrees
1	180 degrees

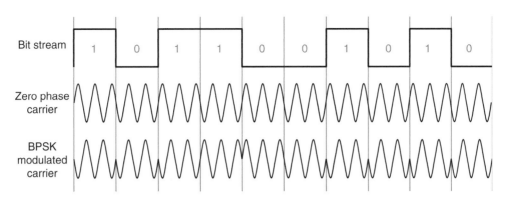

Figure 4-23: Binary Phase Shift Keying Modulation (BPSK)

BPSK modulation is used by IEEE 802.11b at a data rate of 1 Mbps, and by IEEE 802.11a, in combination with OFDM, to achieve data rates of 6 and 9 Mbps.

Quadrature Phase Shift Keying
Instead of the two phase states used in BPSK, QPSK uses four distinct carrier phases, each of which is used to encode a symbol comprised of two input bits or chips.

These four carrier phases are illustrated in Figure 4-24, which represents the phase of the carrier signal in the IQ plane (I = In phase, Q = Quadrature or 90 degrees out of phase). The angle of a given point

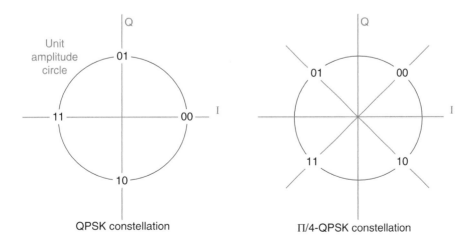

Figure 4-24: QPSK Phase Constellation

Table 4-11: Quadrature Phase Shift Keying

Symbol	Carrier phase
00	0 degrees
01	90 degrees
11	180 degrees
10	270 degrees

on the plane from the I axis represents the phase angle and the distance of a point from the origin represents the signal amplitude. The four points 00, 01, 11, and 10 shown in Table 4-11, are known as the modulation constellation, and represent the four carrier phases each with unit amplitude.

QPSK modulation is used by IEEE 802.11b at a data rate of 2 Mbps, and by IEEE 802.11a, in combination with OFDM, to achieve data rates of 12 and 18 Mbps. π/4-QPSK is a variant of QPSK, as shown in Figure 4-24, which uses carrier phases offset by 45 degrees (i.e. 45, 135, 225 and 315 degrees).

Offset QPSK (O-QPSK) is another variation on QPSK in which transmission of the quadrature phase is delayed by half a symbol period. The consequence is that, unlike QPSK, carrier phase transitions can never be more than 90 degrees and, as a result, the carrier phase and amplitude never passes through zero. The advantage is a narrower spectral width, which is important in applications where interference between adjacent channels must be avoided.

O-QPSK is part of the IEEE 802.15.4 radio specification used by ZigBee, where 16 channels, each 5 MHz wide, are used in the 2.4 GHz ISM band to enable 16 co-located networks. Use of O-QPSK helps to reduce interference between these closely spaced channels.

Differential Phase Shift Keying
Differential phase shift keying is a variation on BPSK and QPSK in which the input symbol results in a differential change of phase instead of defining the absolute phase of the carrier. With BPSK, a 0-symbol corresponds to a period of zero phase carrier, while in DBPSK, a 0-symbol corresponds to no change in carrier phase from the previous bit-period.

Table 4-12: Differential Quadrature Phase Shift Keying

Symbol	*Phase change*
00	0 degrees
01	90 degrees
11	180 degrees
10	270 degrees

Similarly in DQPSK, each symbol translates to a change of phase rather than an absolute carrier phase, as shown in Table 4-12.

Although BPSK or QPSK are conceptually simpler, differential phase shift keying, whether DBPSK or DQPSK, has the practical advantage that the receiver only needs to detect relative changes in carrier phase. The BPSK or QPSK receiver always needs to know the absolute phase reference of the carrier and this reference can be difficult to maintain, for example, if the phase of the received signal is varying due to multipath interference.

Other variants on PSK and DPSK include 8-DPSK, which extends the DQPSK keying table to encode 8 data symbols using phase changes separated by 45 degrees rather than by 90 degrees, and $\pi/4$-DQPSK which, by analogy with $\pi/4$-QPSK, uses carrier phase changes similar to Table 4-12 but offset by 45 degrees (i.e. 45, 135, 225 and 315 degrees). $\pi/4$-DQPSK and 8-DPSK are used in the enhanced data rate (EDR) Bluetooth 2.0 radio for 2 Mbps and 3 Mbps data rates respectively.

Frequency Shift Keying

Frequency shift keying (FSK) is a simple frequency modulation method in which data symbols correspond to different carrier frequencies, as shown in Table 4-13 for BFSK.

Table 4-13: Binary Frequency Shift Keying

Symbol	*Carrier frequency*
0	$f_0 - f_1$
1	$f_0 + f_1$

The sudden carrier waveform changes in simple FSK generate significant out-of-band frequencies and, as a result, FSK is inefficient in terms of spectrum usage. This situation can be improved by passing the input bit stream through a filter to make the frequency transitions more gradual. A Gaussian filter is one type of filter with a specific mathematical form, and use of this as a pre-modulation filter results in Gaussian frequency shift keying (GFSK).

GFSK is used in the Bluetooth radio for standard data rate transmission, with a carrier frequency f_0 of 2.40 to 2.48 GHz and frequency deviation f_1 of between 145 and 175 kHz. Spectral efficiency is particularly important as the FHSS frequency hopping channels are only separated by 1 MHz.

Quadrature Amplitude Modulation

Quadrature amplitude modulation (QAM) is a composite modulation technique that combines both phase modulation and amplitude modulation.

In BPSK or QPSK, a constant carrier amplitude with 2 or 4 different phases is used to represent the input data symbols, as described above. Instead of using 2 or 4 points, QAM defines a constellation of 16, 64 or more points, each with a particular phase and amplitude, and each representing a 4- or 6-bit (or chip) data symbol.

16-QAM and 64-QAM modulation techniques are used in the IEEE 802.11a and g specification, together with OFDM, to achieve data rates of 24 to 54 Mbps. Figure 4-25 shows the 16-QAM constellation — the 16 points on the IQ plane — used to achieve data rates of 24 Mbps and 36 Mbps.

The points in the 16-QAM constellation can be alternatively numbered according to a Gray coding in which adjacent points differ only in the switching of one bit, as shown in Figure 4-26. Using this numbering reduces the chance of two-bit errors in the receiver — if a point is erroneously detected as a neighbouring point only one bit will be incorrect. This makes it easier to recover the bit error using error correction techniques.

The next step, a 256-QAM modulation scheme, would further improve achievable data rate with no increase in the occupied bandwidth, but

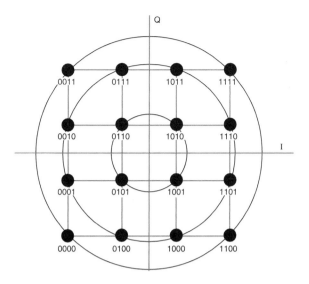

Figure 4-25: 16-QAM Constellation

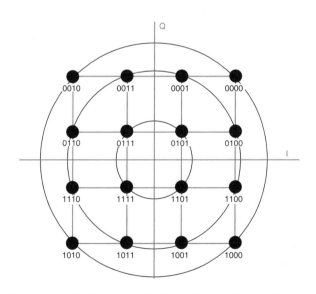

Figure 4-26: Gray Coded 16-QAM Constellation

generating and processing 256-QAM modulated signals is currently a significant challenge for hardware performance and cost.

Dual Carrier Modulation

Dual carrier modulation is a technique applied in multi-carrier systems such as OFDM to combat the loss of data due to the destructive interference, or fading, of individual carrier signals in a multi-path environment (see the Section "Multipath Fading, p. 114"). By modulating data onto two carrier frequencies rather than one, the transmission can be made more robust, although at the cost of using additional bandwidth.

In multi-band OFDM (see the Section "Multiband UWB, p. 122") a 4-bit symbol is mapped onto two different 16-QAM constellations and the resulting symbols are transmitted on two OFDM carrier tones separated by at least 200 MHz. If reception of one of the tones is affected by fading, the data can be recovered from the other tone, the wide separation assuring that the probability of both tones being affected is very small.

Pulse Modulation Methods

Several wireless network standards specify the use of pulsed rather than continuous transmission of a carrier wave, and a number of specific modulation techniques are used in these systems.

Pulse Position Modulation

In pulse position modulation (PPM), each pulse is transmitted within a reference time frame, and the information carried by the pulse is determined by the specific transmission time of the pulse within its frame. For example, a 4-PPM system will define four possible positions for a pulse within the reference frame, with each possible position coding one of four input data symbols (Table 4-14).

More generally, an *m*-PPM system will have *m* possible pulse transmission slots within a frame. An 8-PPM modulation system is shown in Figure 4-27. PPM is specified in the IrDA standard at a 4 Mbps data rate, and is also used in Impulse Radio (IR as opposed to Ir!).

Table 4-14: Data Symbols for 4-PPM Modulation

Input data symbol	*4-PPM data symbol*
00	1000
01	0100
10	0010
11	0001

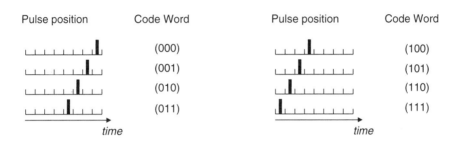

Figure 4-27: 8-PPM Modulation

Pulse Shape Modulation

Pulse shape modulation (PSM) encodes the input data stream in the shape of the transmitted pulse. The simplest form of PSM is pulse amplitude modulation (PAM) in which, typically, two or four distinct pulse amplitudes are used to encode data symbols, as shown in Table 4-15.

Similarly, pulse width modulation (PWM) uses the width of transmitted pulses and more generally, PSM may use some other pulse shape characteristics — such as the derivative of the pulse waveform — to encode data onto the pulse train.

Pulse amplitude and pulse shape modulation are candidates for use in the ultra wideband radio physical layer of the ZigBee specification.

Table 4-15: PAM Encoding Table

Input data symbol	Pulse amplitude
00	0
01	1
10	2
11	3

RF Signal Propagation and Reception

The first part of this chapter described the various techniques that are used to encode and modulate an input data stream onto a radio frequency carrier. The following four sections discuss the various factors that impact on the transmission, propagation and reception of radio waves, which will enable an estimate to be made of the power requirements for a given wireless networking application.

The key factors are transmitter power, antenna gain at the transmitter, propagation or link losses, antenna gain once again at the receiving station, and finally receiver sensitivity. Taken together these factors make up the link budget — the balance of power plus gain required to compensate for losses in the link so that sufficient signal strength is available at the receiver to allow data decoding at an acceptable error rate.

Transmitter Power

Every RF transmitter generates a certain amount of power (P_{TX}), which is the first major factor in determining the range of a radiated signal. Transmitter power is measured in one of two ways, either in the familiar unit of Watts (or milliwatts) or alternatively using a relative unit called "dBm". The power in dBm is calculated as dBm = $10 \times \log_{10}$ (Power in milliwatts), so a transmitter of 100 mW (0.1 Watts) is equivalent to 20.0 dBm (Table 4-16).

The dB (or dBm) unit is useful for two reasons. First, when considering the various factors affecting signal strength, these effects can be easily combined when using dB units by simply adding the relevant dB numbers together. Second, it is easy to translate dB into relative power by remembering that +3 dB represents a doubling of power and –3 dB similarly a halving of power. The additive rule applies here too, so –6 dB is ¼ the power, –9 dB is ⅛ and so on.

Table 4-16: Power in mW and dBm

Power (mW)	Power (dBm)
0.01	−20
0.1	−10
0.5	−3
1	0
10	10
20	13
100	20
1000	30

Transmitter power levels for typical wireless networking products are in the region of 100 milliwatts to 1 Watt (20 to 30 dBm). For example, in the US the FCC specifies a maximum transmitter power of 1 Watt for FHSS and DSSS transmitters in the 2.4 and 5.8 GHz ISM bands. In the UK the Radio Communications Agency (RA) specifies that these devices must have a maximum effective isotropic radiated power (EIRP) in the 2.4 GHz band of 100 mW or 20 dBm. As described below, EIRP is a combination of the transmitter's power and the antenna gain.

Antenna Gain

An antenna converts the power from the transmitter into electro-magnetic waves that are radiated to the receiver, and the type of antenna affects the pattern and power density of this radiation, and therefore the strength of signal seen by a receiver. For example, a simple dipole antenna emits radiation relatively evenly in all directions apart from along its axis, while a directional antenna emits radio waves in a narrow beam. Typical radiation patterns of a dipole and a directional antenna are shown in Figure 4-28.

The ratio of the maximum power density at the centre of the radiation pattern of any antenna to the power density of the radiation from a reference isotropic antenna is known as the antenna gain (G_{TX} or G_{RX}), and is measured in dBi units. The effective isotropic radiated power, or EIRP, of a radiating antenna is then, the sum of the dBm power arriving at the antenna from the transmitter plus the dBi antenna gain. The types of antenna that can be used in wireless

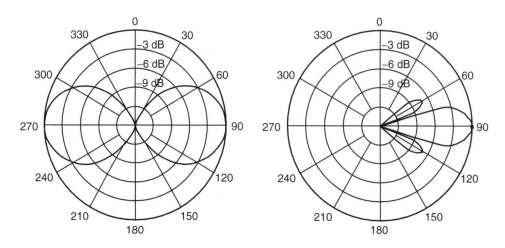

Figure 4-28: Radiation Pattern from Dipole and Yagi Antennas

networks were described in the Section "Wireless LAN Antennas, p. 55" which illustrated antennas with gain ranging from 0 dBi for an omnidirectional dipole antenna to +20 dBi or more for a narrow beam directional antenna.

The cables and connectors that link the transmitter or receiver to the antenna also introduce a loss into the system that can range from a few dB to tens of dB depending primarily on the length and quality of cabling. An equally important aspect of transmitter to antenna or receiver to antenna connections, is the matching of impedance between these components. For example, maximum power will only be transmitted if the impedances of the transmitter, connecting cable and antenna are equal, otherwise power will be lost as a result of reflections at the connections between components. In the case of equipment with integrated antennas, this will be part of the design dealt with by the equipment manufacturer. However, this aspect will have to be considered when attaching an external antenna to a wireless network adapter or access point.

Receiver Sensitivity

As the strength of the signal reaching the receiver input drops, the decoding of data will be increasingly affected by noise and, as a result, will become increasingly error prone. The sensitivity limit of the receiver is determined by the allowable bit error rate and the receiver noise floor.

As these factors are discussed in the following sections, an example will be worked through, based on the following parameters:

- 802.11b DSSS system (2.4 GHz, 22 MHz spread bandwidth)

- DQPSK modulation

- 2 Mbps data rate or 2 MHz de-spread bandwidth

- required bit error rate of 1 in 10^5

- receiver noise figure of 6 dB

- 20°C ambient temperature.

Bit Error Rate

The rate at which decoding errors occur is measured by the bit error rate (*BER*), with a *BER* of 1 in 10^5 being typical at the receiver sensitivity limit. Since data is transmitted in packets containing several hundreds or thousands of bits of data, even a 1 in 10^5 chance of an error in decoding any single bit will multiply up to a significant probability of an error in a large data packet, and the resulting packet error rate (*PER*) can be in the range of several percent. For example, with a *BER* of 1 in 10^5 the *PER* for a 100 bit data packet will be:

$$(1 - PER) = (1 - 10^{-5})^{100}$$

or *PER* = 0.1%, rising to 1% for a 1 kb data packet.

The *BER* is a function of the signal-to-noise ratio in the receiver, and also depends on the specific type of modulation method being used. The signal-to-noise ratio of a communication channel is given by:

$$SNR = (E_b / N_o) * (f_b / W) \qquad (4.1)$$

where E_b is the energy required per bit of information (Joules), f_b is the bit rate (Hz), N_o is the noise power density (Watts/Hz) and W is the bandwidth of the modulated carrier signal (Hz).

Note that for our example, considering a DSSS system, it is the de-spread bandwidth that is considered in Eq. 4.1. Using a spread spectrum rather than a narrow band transmission results in an additional gain known as the processing gain.

$$\text{Processing Gain} = 10 \log_{10}(C) \text{ dB} \qquad (4.2)$$

where C is the code length in chips (11 for the Barker code discussed above). This processing gain is effectively included in the calculation of channel *SNR* by using the de-spread bandwidth in Eq. 4.1.

The bit rate per Hz of bandwidth, f_b/W, is a function of the modulation method employed, as discussed in the Section "Introduction, p. 95". *BER* is then given by:

$$BER = \tfrac{1}{2}\, erfc\, (SNR)^{1/2} \tag{4.3}$$

where *erfc* is the so called complementary error function which can be looked up in mathematical tables. Figure 4-29 shows *BER* as a function of *SNR* for some of the common modulation methods.

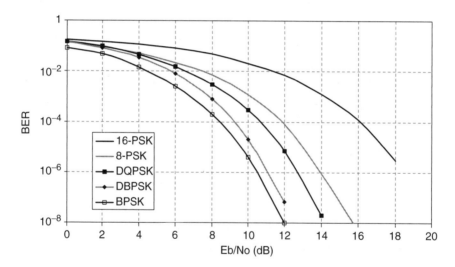

Figure 4-29: Bit Error Rate (*BER*) for Some Common Modulation Methods

The figure shows that for the example DQPSK modulated signal, with 1 bit per Hz of bandwidth, a signal-to-noise ratio of 10.4 dB is required to achieve a bit error rate of 1 in 10^5.

Receiver Noise Floor

The receiver noise floor has two components, the theoretical thermal noise floor (N) for an ideal receiver, and the receiver noise figure (NF)

which is a measure of the additional noise and losses in a particular receiver. The thermal noise is given as:

$$N = kTW \qquad (4.4)$$

where k is the Boltzmann constant (1.38×10^{-23} Joules/°K), T is the ambient temperature in °K and W is the bandwidth of the transmission (Hz)[1]. Receivers for wireless networking will typically have noise figures in the range from 6 to 15 dB.

The receiver noise floor (RNF) is then the sum of these two terms:

$$RNF = kTW + NF \qquad (4.5)$$

For the example 802.11b receiver with a 2 MHz de-spread bandwidth, operating at 20°C (290°K) and with a noise figure of 10 dB:

$$N = 1.38 \times 10^{-23} \text{ J/K} \times 290°K \times 2 \times 10^6 \text{ Hz}$$

$$= 8.8 \times 10^{-12} \text{ mW}$$

$$= -110.6 \text{ dBm}$$

$$RNF = -110.6 \text{ dBm} + 10 \text{ dB}$$

$$= -100.6 \text{ dBm}$$

Receiver Sensitivity

The receiver sensitivity, P_{RX}, is the sum of the receiver noise floor (RNF) and the signal-to-noise ratio (SNR) required to achieve the desired bit error rate:

$$P_{RX} = RNF + SNR \qquad (4.6)$$

For the example;

$$P_{RX} = -100.6 + 10.4 \text{ dBm}$$

$$= -90.2 \text{ dBm}$$

From this discussion it can be seen that as the data rate in the example increases from 2 Mbps towards the 802.11b maximum of 11 Mbps, different modulation methods will be needed to achieve the higher bandwidth efficiency (more bits per Hz of bandwidth, or f_b/W in Eq. 4.1).

---

[1]Again, the de-spread bandwidth is used here.

Table 4-17: P_{RX} Versus Data Rate for a Typically 802.11b Receiver

Data Rate (Mbps)	Modulation technique	P_{RX} (dBm)
11	256 CCK + DQPSK	−85
5.5	16 CCK + DQPSK	−88
2	Barker + DQBSK	−89
1	Barker + DBPSK	−92

This will result in a higher signal-to-noise ratio requirements for the same bit error rate, so that the receiver sensitivity will decrease at higher data rates. This is shown in Table 4-17 for a typical 802.11b receiver.

This dependence of P_{RX} on data rate underlies the gradual deterioration in wireless network throughput as signal strength decreases. There is no abrupt cut-off in performance, but rather a gradual reduction in throughput as the transmitter and receiver switch to a lower data rate at which a low *BER* can be maintained.

RF Signal Propagation and Losses

Between the transmitting and receiving antennas, the RF signal is subject to a number of factors that affect signal strength. These are considered in the following sections.

Free Space Loss

Once the signal is radiating outwards from the antenna, the signal power falls off with distance due to the spreading out of the radio waves. This is known as free space loss, and overall is the most significant factor affecting received signal strength.

Free space loss is measured in dB, and depends on the signal frequency and transmission distance according to the formula:

$$L_{FS} = 20 \log_{10} (4\pi D / \lambda) \tag{4.7}$$

where D is the transmitter to receiver distance in metres. λ is the wavelength of the radio signal in metres which can also be expressed as:

$$\lambda = c / f \tag{4.8}$$

where c is the speed of light (3×10^8 m/s) and f is the signal frequency in Hz.

Expressing these quantities in more convenient units, f in MHz and D in km, L_{FS} can be calculated from the formula:

$$L_{FS} = 32.45 + 20 \log_{10}(f) + 20 \log_{10}(D) \qquad (4.9)$$

So, at 2.4 GHz (2400 MHz) the free space loss at 100 m (0.1 km) will be:

$$L_{FS} = 32.4 + 20 \log_{10}(2400) + 20 \log_{10}(0.1)$$
$$= 32.4 + 67.6 - 20$$
$$= 80 \text{ dB}$$

The third term in Eq. 4.9 shows that L_{FS} increases by 20 dB for every factor of 10 increase in range. This gives the useful rule of thumb for the 802.11b 2.4 GHz band that L_{FS} is 60 dB at 10 metres, 80 dB at 100 metres and so on.

From the second term in Eq. 4.9 it can be seen that the frequency dependence results in an increase in L_{FS} for transmissions at 5.8 GHz of $20 \log_{10}(5800/2400)$ or 7.7 dB (see Figure 4-30). While this may be important in open-air applications, in practical indoor situations, this difference is often small compared with other environmental effects.

This calculation of free space loss assumes that there is a clear line-of-sight between the transmitting and receiving antennas, which means that the receiver can effectively "see" the transmitter. However, to maximise RF propagation range in the open-air it is not enough just to be able to see in

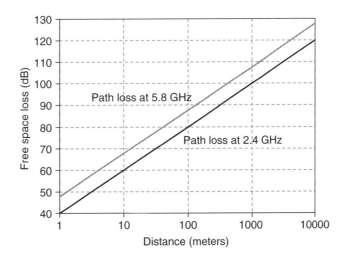

Figure 4-30: Free Space Loss at 2.4 GHz and 5.8 GHz

a straight line between the two antennas. The volume of space around this straight line affects signal propagation as well, and any obstructions that come close to the direct line-of-sight will also cause signal loss.

Fresnel Zone Theory

The theory used to calculate the effect of obstructions is called Fresnel zone theory. The Fresnel zone is a region between the two antennas with an oval shape similar to a rugby ball.

There are actually a series of such regions, called the 1st, 2nd, 3rd, etc. Fresnel zones (Figure 4-31), and at the mid point between transmitter and receiver, the radius of the *n*th Fresnel zone in metres is calculated from the formula:

$$R = 0.5 \, (n \times \lambda \times D) \tag{4.10}$$

where the wavelength and the transmitter to receiver distance are also in metres. For a 2.4 GHz signal, with a wavelength of 12.5 cm (0.125 m), the first Fresnel zone has a mid-point radius of 1.8 m for a 100 m range, or 5.6 m for a 1 km range. Any obstructions within the first Fresnel zone will cause signal loss through reflection, refraction or diffraction.

Multipath Fading

Multipath fading occurs when reflected, refracted or diffracted signals travel to the receiver along different paths, resulting in a range of different arrival times known as the multipath delay spread. Signals arriving along different paths will be phase-shifted with respect to the direct path signal,

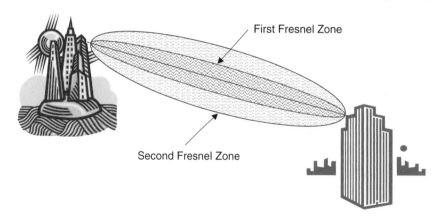

Figure 4-31: The Fresnel Zones Around a Propagation Path

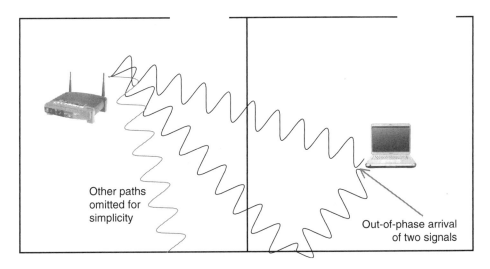

Other paths
omitted for
simplicity

Out-of-phase arrival
of two signals

Figure 4-32: Multi-path Fading in an Indoor Environment

as shown in Figure 4-32, and will therefore cause some degree of destructive interference at the receiving antenna. This is familiar in UHF TV reception as a ghost image caused by an interfering signal reflected from a nearby building or other large object.

Interference between these multiple delayed signals can substantially reduce the signal strength at the receiving antenna, introducing a loss that can be as much as 20 to 30 dB.

It is possible to compute multipath losses using complex ray tracing or other algorithms, but this is rarely done in practice.

Signal Attenuation Indoors
For a typical wireless network in a home or small office, multiple obstructions such as walls, floors, furniture and other objects will obstruct the propagation path from transmitter to receiver, and signal reception will tend to be very variable. Depending on its construction, transmission through a wall can introduce a loss of 3 to 6 dB or more, as shown in Table 4-18, and an additional allowance will be required in the link budget to account for this loss.

In a multi-storey building, losses between floors will also depend on the building materials used, and will be very high in buildings

Table 4-18: Typical Attenuation for Building Materials at 2.4 GHz

Attenuation range	*Materials*	*Loss (dB)*
Low	Non tinted glass, wooden door, cinder block wall, plaster.	2–4
Medium	Brick wall, marble, wire mesh or metal tinted glass.	5–8
High	Concrete wall, paper, ceramic bullet-proof glass.	10–15
Very high	Metal, silvering (mirrors).	>15

with sheet steel construction. More typically, a loss of approximately 6 dB is seen between adjacent floors, rising to around 10 dB per additional floor for separations of two to three floors. The typical losses shown in Table 4-18 are highly dependent on the specific materials used and methods of construction. For example, even a stud wall can introduce a significant loss if it contains a fire retarding foil membrane.

In common with multi-path fading, complex algorithms are required to calculate the various types of losses indoors, and it is therefore convenient to combine these loss components to give a single additional term in the link budget. This term, the fade margin (L_{FM}), will generally be estimated by a rule-of-thumb, or determined by an on-site survey.

Link Budget

The factors considered above, transmitter power (P_{TX}), antenna gain at the transmitter (G_{TX}) and receiver (G_{RX}), receiver sensitivity (P_{RX}), free-space loss (L_{FS}) and other losses combined in the fade margin (L_{FM}), together

Table 4-19: Balancing factors in the link budget

Reducing required P_{TX}	*Increasing required P_{TX}*
Lower receiver sensitivity (bigger negative) P_{RX}	Higher free space loss L_{FS}
Higher transmitter antenna gain G_{TX}	Higher fade margin L_{FM}
Higher transmitter antenna gain G_{RX}	

define the link budget that is available to bring the data signal successfully from transmission to detection (Table 4-19).

It is convenient to express the link budget in terms of the transmitter power (P_{TX}) required to deliver a signal to the receiver at its sensitivity limit (P_{RX}). Expressed in dBm, this is:

$$P_{TX} = P_{RX} - G_{TX} - G_{RX} - L_{FS} - L_{FM} \quad \text{dBm} \tag{4.11}$$

For example, a system comprising a directional transmitting antenna with a gain of 14 dBi, a patch receiving antenna (6 dBi), and a receiver with a sensitivity of −90 dBm, operating over 100 metres at 2.4 GHz (L_{FS} = 80 dB) with a 36 dB fade margin (L_{FM}) results in a required transmitter power of:

$$P_{TX} = -90 \text{ dBm} - 14 \text{ dBi} - 6 \text{ dBi} + 80 \text{ dB} + 36 \text{ dBm}$$
$$= +6 \text{ dBm}$$

To ensure that the signal at the receiving antenna is above the receiver sensitivity, the required transmitter power is therefore +6 dBm (4 mW), as shown graphically in Figure 4-33. This configuration would be comfortably achieved with a 100 mW (20 dBm) transmitter, with an extra 14 dB link margin for unaccounted losses or noise.

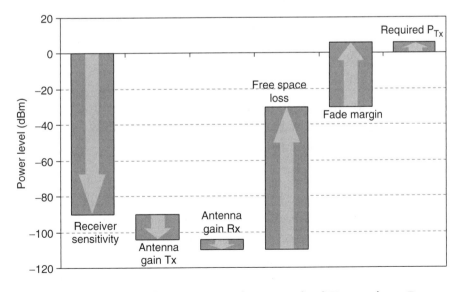

Figure 4-33: Link Budget Expressed as Required Transmitter Power

Ambient Noise Environment

As well as the receiver noise floor, which defines the limit of receiver sensitivity, other sources of external RF noise will also have an impact on the reliability of RF signal detection and data decoding.

The total RF noise entering a radio antenna at any particular location is termed the ambient noise environment and is made up of two components:

■ Ambient noise floor — the aggregate background noise from distant sources such as car ignition, power distribution and transmission systems, industrial equipment, consumer products, distant electrical storms and cosmic sources.

■ Incidental noise — the aggregate background noise from localised man-made sources.

The ambient noise floor is generally "white noise", with constant power per unit bandwidth, while incidental noise may be either broadband or narrow band. In implementing local or metropolitan area wireless networks, the ambient noise floor will be established during an RF site survey, and should be explicitly included in the link budget if it is above the receiver sensitivity of planned equipment.

For example, in the link budget calculation above, if an ambient noise floor of –85 dBm was measured, using this in place of the receiver sensitivity of –90 dBm would result in a required transmitter power of 11 dBm. This type of environmental noise will limit the range that can be achieved for a given equipment configuration (transmitter power, antenna gains, etc.), but will not degrade the performance of wireless networks when operating within that limit.

Narrow band incidental noise, from nearby man-made sources such as a microwave oven or a narrow band transmitter, is more likely to result in unpredictable and unreliable network performance.

Interference Mitigation Techniques

Wireless networking specifications are increasingly including a range of measures to mitigate the effect of interference on network performance. Wireless USB, covered in Chapter 10, is a good example of the approach which starts with establishing information about the quality of the RF link and then provides measures to control various link characteristics.

Table 4-20: Wireless USB Interference Mitigation Controls

Control	*Description*
Transmit power (TPC)	Host can control its own transmit power level as well as querying and controlling transmit power of devices in the cluster.
Transmitted bit rate	Host can adjust the transmitted bit rate for both outward (host to device) and inward (device to host) transfers.
Data payload size	When interference causes *PER* to rise, reducing packet size can improve throughput by reducing uncorrectable errors.
RF channel selection	Wireless USB's MB–OFDM radio provides multiple alternative channels which can be used by a host if supported by all devices in the cluster.
Host schedule control	Allowing isochronous data transfers to temporarily use channel time allocated for asynchronous transfers, in order to retransmit failed isochronous data packets.
Dynamic bandwidth control	Host control of the spectral shaping capabilities of the MB–OFDM UWB radio, described in the following section.

In wireless USB, the host and other devices can maintain statistical information on packet error rate and on link indicators such as received signal strength (RSSI) and link quality (LQI). The latter indicator measures the error in the received modulation of successfully decoded symbols.

The main RF link controls available in wireless USB are described in Table 4-20.

Transmit power control and RF channel selection are also included in the network optimisation measures introduced in the 802.11k extension to the Wi-Fi networking standard, covered in Chapter 6.

Ultra Wideband Radio

Introduction

Ultra Wideband wireless communication systems are based on impulse radar technology that was developed for military applications by the

USA and USSR in the 1960s. Impulse radar or radio transmits extremely short electromagnetic pulses, typically less than 1 ns (nanosecond) in length, with no underlying carrier signal. Such short pulses result in an effective bandwidth of the transmission that may be from 500 MHz up to several GHz.

In 2002 the FCC in the USA opened 7.5 GHz of radio spectrum for UWB applications, from 3.1 to 10.6 GHz, and adopted a definition of UWB as any intentional transmission in which the bandwidth to −3 dB points was at least 20% of the mean frequency of the transmission, with a minimum bandwidth of 500 MHz.

Since UWB transmissions cover a wide swath of the radio spectrum, an important requirement is that they do not result in harmful interference with other RF transmitted services, whether current or planned. To ensure this coexistence, the FCC has defined strict EIRP limits on UWB transmission, as shown in Figure 4-34. The maximum permitted power density (EIRP) of −41.3 dBm/MHz is below the FCC Part 15 noise power limit for unintentional emitters such as computers and other electronic devices. As a result of this very low EIRP specification, UWB wireless is suited for applications where very long battery life is required.

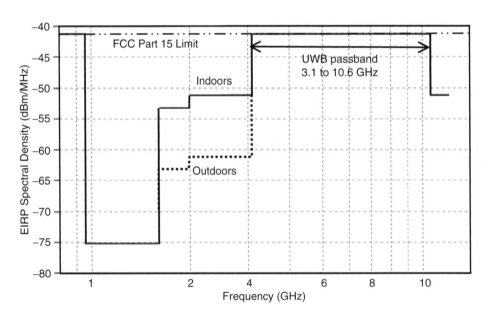

Figure 4-34: FCC UWB Passband Specification

A second characteristic of some UWB implementations is spectral shaping — the capability to control the radiated power spectrum in order to avoid transmission at particular narrow band frequencies.

UWB comes in three varieties for data communication applications:

- time-hopping pulse position modulation (or Impulse radio)

- direct sequence spread spectrum–UWB (DS–UWB)

- multiband–UWB (such as multiband OFDM).

Of these, MB–OFDM offers the greatest flexibility in spectral shaping, with a wide range of course and fine control options easily implemented in software.

Time Hopping PPM UWB (Impulse Radio)

Impulse radio (IR) is the name given to UWB radio based on time-hopping, pulse position modulation. Data is transmitted as a discontinuous series of very short pulses, with one pulse per user in each time hopping frame of length T_f. The nominal transmission time of a pulse in a given frame is determined by a pseudo-noise (PN) code that is specific for each user of the communication channel.

Finally, whether a pulse represents a 1-bit or a 0-bit depends on the actual transmission time relative to the nominal transmission time (the pulse position modulation). For example, in an early/late PPM system, if the pulse is transmitted a time offset δ ahead of the nominal time it represents a 1-bit, or if an offset δ after the nominal time then it represents a 0-bit.

In the example shown in Figure 4-35, a TH code of length 4 is used, so that four pulses are transmitted for each bit, each pulse in one of the eight code slots (T_c) in each of four successive frames.

Pulse amplitude modulation (PAM) or pulse shape modulation (PSM) can be used as alternatives to PPM, with 1-bit and 0-bit then being determined by the amplitude or shape of each individual pulse.

Impulse radio is one of two optional physical layer specifications selected in March 2005 by the IEEE 802.15 Task Group 4a as part of the enhancement of the 802.15.4 standard. (The other optional PHY is a chirp spread spectrum operating in the 2.4 GHz ISM band.)

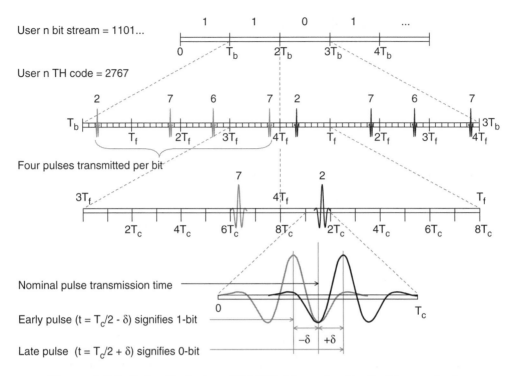

Figure 4-35: Pulse Train in a TH-PPM Impulse Radio Transmission

Direct sequence UWB (DS-UWB)

Direct sequence, discussed in the Section "Direct Sequence Spread Spectrum in the 2.4 GHz ISM Band, p. 83" as the spread spectrum technique underlying the IEEE 802.11b and 802.11g physical layer, can also be applied in UWB radios. Instead of the chipping code being used to spread the carrier spectrum by increasing the symbol transmission rate, the spectrum is spread to UWB proportions as a result of the very narrow pulse that is used to transmit each symbol.

The chipping code then plays a multiple access role (CDMA) with individual user codes determining the exact times at which individual users of the channel will transmit or receive a pulse. A variety of modulation methods (PAM, PSM) can be used to code the data stream onto the pulse stream. So far, DS-UWB has not been identified as a target technology for any wireless networking applications.

Multiband UWB

In multiband UWB, ultra-wide bandwidth is achieved by dividing the frequency band of interest into multiple overlapping or adjacent bands,

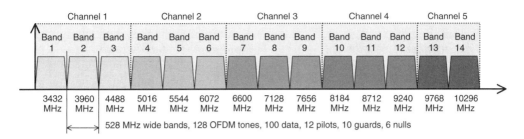

Figure 4-36: MB–OFDM Frequency Bands and Channels

and operating simultaneously on all available bands. Currently, the most important example of this technique is multiband (MB) OFDM, which is being promoted by the MB–OFDM Alliance (MBOA) and has been adopted as the basis for Wireless USB and Wireless FireWire (Wireless 1394).

MB–OFDM, as proposed by the MBOA, uses a bandwidth from 3.168 GHz to 10.560 GHz, which is divided into 14 bands of 528 MHz full width — thus meeting the FCC's 500 MHz minimum bandwidth specification. The 14 bands are grouped into 5 band groups or channels, as shown in Figure 4-36.

Frequency hopping between bands within a band group can be used to enable overlapping piconets to be formed but unlike Bluetooth, which makes 1600 hops per second across 79 frequencies, the MBOA radio as specified for wireless USB makes 3 million hops per second, one hop after every transmitted symbol, across just 3 frequencies.

MBOA specifies two types of time-frequency codes (TFC) as shown in Table 4-21. Time-frequency interleaving (TFI) codes define frequency hopping patterns, while fixed frequency interleaving (FFI) codes define continuous transmission on a single OFDM band. The FFI option can be used to improve the performance of two or more simultaneously operating piconets, by assigning a single OFDM band to each piconet.

Within each 528 MHz band, 128 OFDM subcarriers are transmitted, with data modulated onto 100 of these and the remainder used as pilot, guard and null tones. For data rates of up to 200 Mbps, MBOA specifies data modulation using QPSK, while rates of 320 to 480 Mbps use dual carrier modulation (DCM).

Table 4-21: MBOA Time-frequency Codes

Code number	Code type	Band number (Band group 1)					
1	TFI	1	2	3	1	2	3
2	TFI	1	3	2	1	3	2
3	TFI	1	1	2	2	3	3
4	TFI	1	1	3	3	2	2
5	FFI	1	1	1	1	1	1
6	FFI	2	2	2	2	2	2
7	FFI	3	3	3	3	3	3

Spectral shaping is used to avoid interference with other RF services and can be changed under software control to respond to specific local regulations or time-varying conditions. Coarse control can be achieved by dropping whole bands (or in extreme circumstances whole band groups), but extremely precise shaping is also possible by "nulling out" a certain number of tones within a single band.

MIMO Radio

As described in the Section "Multipath Fading, p. 114", the multiple paths that a radio signal takes between transmitter and receiver often lead to a degradation of signal strength through multi-path fading. Multi-input multi-output (MIMO) radio takes advantage of this characteristic of RF propagation by sending multiple data streams across multiple transmitters to receiver paths in order to achieve a higher data capacity (Figure 4-37). Mathematical modelling of the propagation paths, using a channel calibration period during each transmitted data packet, allows the different signal paths and data streams to be identified and correctly recombined in the receiver.

This technique, space division multiplexing (SDM), is analogous to FDM in the frequency domain but instead of different frequencies carrying data in parallel, here different spatial paths carry data in parallel.

Effectively the same bandwidth is being used simultaneously to create multiple communication paths. If these paths are equally strong and can

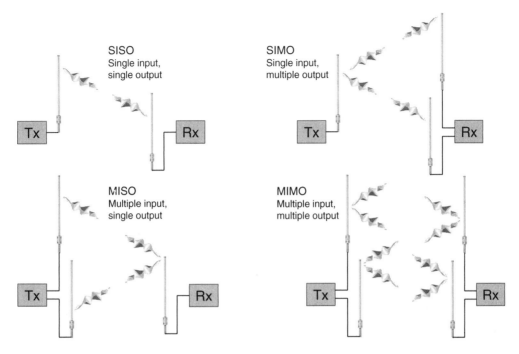

Figure 4-37: MIMO Radio Definition

be perfectly separated, the overall capacity of the communication channel increases linearly with the number of independent paths used. In a system with M transmitters and N receivers, the number of independent paths is the minimum of M and N.

In practice, all paths will not be equally strong or perfectly separated, and performance will be determined by coefficients, known as singular values, which characterise each path between a transmitting and receiving antenna. These singular values are determined by including a short "training period" in the preamble of each transmitted data packet, during which known and different signals are transmitted from each antenna. These signals provide information about the transmission channel (so-called Channel State Information or CSI), and with this information the receiver can compute the singular values that are used to decode the remainder of the data packet.

The increased capacity of MIMO radio can be used to achieve a higher data rate or to increase link robustness or range for a given data rate. The IEEE 802.11n specification (described in the Section "MIMO and Data

Rates to 600 Mbps (802.11n), p. 165") will use MIMO to increase the PHY layer data capacity of the 802.11a/g radio from 54 Mbps to in excess of 200 Mbps.

Space time block coding (STBC) is a related technique which combines space and time diversity to increase the robustness or range of an RF link. STBC breaks the transmitted data into blocks and transmits multiple time-shifted copies of each block of data from each transmitting antenna to the receiving antenna. STBC is thus a Multi-Input Single-Output (MISO) technique (see Figure 4-37), although multiple receiving antennas can further improve performance.

Near Field Communications

Introduction
Near field communications (NFC) is a very short range radio frequency communications technology that has been extensively developed for use in RF identification (RFID) tags and other smart labelling applications. These applications have typically employed a RF carrier frequency of 13.56 MHz, which is internationally allocated as an unlicensed ISM band.

NFC is distinct from so-called far field RF communication used in personal area and longer range wireless networks, since it relies on direct magnetic field coupling between transmitting and receiving devices.

There are two types of NFC devices, active and passive, which operate quite differently. Passive devices do not have an internal power source, but derive their power from an active initiating device by inductive coupling. A passive device also does not transmit data by generating a magnetic field as an active device does. Instead a passive device transfers data back to an active device through a process called load modulation. These concepts are described in the following sections.

Near Field and Far Field Communication
The space around an antenna can be divided into two regions based on the differing nature of the electromagnetic fields generated by the antenna. The boundary between the two regions is known as the radian sphere and has a radius of $\lambda/2\pi$, where λ is the wavelength of the propagated electromagnetic wave.

The primary magnetic field begins at the antenna and oscillations in this field induce an electric field in the surrounding space. This region, inside the radian sphere, is within the influence of the primary magnetic field and is called the near field of the antenna. The electromagnetic field equations in this region reflect energy storage in the magnetic field and are described by near field coupling volume theory.

The region outside the radian sphere is called the far field of the antenna and here the fields separate from the antenna and propagate into space as an electromagnetic wave. The electromagnetic field equations here represent energy propagation rather than storage, and propagation is described in terms of the concepts covered in the Section "RF Signal Propagation and Reception, p. 106".

For NFC operating at 13.56 MHz, $\lambda = 22$ metres, so that the radius of the radian sphere is $\lambda/2\pi = 3.5$ metres. In the near field region, the magnetic field strength is inversely proportional to the cube of the distance between the antennas while the power in the magnetic field, which is used to energise passive NFC, decreases as the inverse sixth power of the separation. This is equivalent to an attenuation of 60 dB for a ten-fold increase in distance.

Inductive Coupling

Near-field inductive coupling uses an oscillating magnetic field to transfer RF energy between devices. Each device includes a resonant circuit tuned to the RF carrier frequency, and a loosely coupled "space transformer" is established when the coil windings or "antenna loops" of the two devices are brought into range (Figure 4-38). The effective range is comparable to the actual physical dimensions of the transmitting antenna loop.

When the resonant circuit in the transmitting device is energised by a RF power source, the resulting magnetic flux linkage results in energy transfer between the two resonant coil windings.

Inductive coupling is only effective in the near-field region of the transmitting antenna loop. In the far-field region, where the electromagnetic field separates from the antenna and propagates as an electromagnetic wave, it can no longer have a direct effect through inductive coupling.

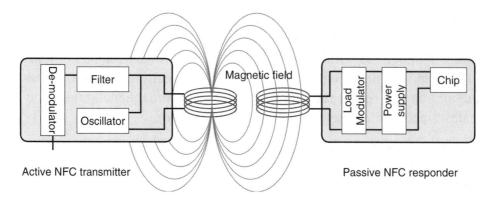

Figure 4-38: Inductive Coupling Between NFC Antenna Loops

Load Modulation

When an NFC target device in passive mode is within range of an active NFC transmitter (or initiator) its resonant circuit draws energy from the magnetic field created by the initiating device. This additional consumption results in a voltage perturbation that can be measured in the resonant circuit of the initiator. If an additional load resistance in the target is periodically switched on and off, this has the effect of an amplitude modulation of the carrier wave voltage in the initiator.

By using the data stream to be transmitted from the target device to control this load switching, the data stream is transferred from the target to the initiator. This technique is called load modulation.

Load modulation creates amplitude modulated sidebands on the 13.56 MHz carrier frequency, and the data stream is recovered by demodulating these sidebands in the initiating device's RF signal processing circuits.

Infrared Communication Basics

The Ir Spectrum

The infrared (Ir) part of the electromagnetic spectrum covers radiation having a wavelength in the range from roughly 0.78 μm to 1000 μm (1 mm). Infrared radiation takes over from extremely high frequency (EHF) at 300 GHz and extends to just below the red end of the visible light spectrum at around 0.76 μm wavelength. Unlike radio frequency radiation, which is transmitted from an antenna when excited by an oscillating electrical signal, infrared radiation is generated by the rotational and vibrational oscillations of molecules.

The infrared spectrum is usually divided into three regions, near, middle and far, where "near" means nearest to visible light (Table 5-1). Although all infrared radiation is invisible to the human eye, far infrared is experienced as thermal, or heat, radiation. Rather than using frequency as an alternative to wavelength, as is commonly done in the RF region, the wavenumber is used instead in the infrared region. This is the reciprocal of the wavelength and is usually expressed as the number of wavelengths per centimetre.

One aspect of wireless communication that becomes simpler outside the RF region is spectrum regulation, since the remit of the FCC and equivalent international agencies runs out at 300 GHz or 1 mm wavelength.

Infrared Propagation and Reception

The near infrared is the region used in data communications, largely as a result of the cheap availability of infrared emitting LEDs and optodetectors,

Table 5-1: Subdivision of the Infrared Spectrum

Infrared region	*Wavelength (μm)*	*Wavenumber (/cm)*
Near	0.78–2.5	12,800–4000
Middle	2.5–50	4000–200
Far	50–1000	200–10

solid state devices that convert an electrical current directly into infrared radiation and vice versa. Infrared LEDs emit at discrete wavelengths in the range from 0.78 to 1.0 μm, the specific wavelength of the LED being determined by the particular molecular oscillation that is used to generate the radiation.

Transmitted Power Density – Radiant Intensity

Ir propagation is generally a simpler topic than RF propagation, although the same principles, such as the concept of a link budget, still apply. As described in Chapter 4, the link budget predicts how much transmitter power is required to enable the received data stream to be decoded at an acceptable BER. For Ir, the link budget calculation is far simpler than for RF, as terms like antenna gain, free space loss and multipath fading no longer apply. As a result, propagation behaviour can be more easily predicted for Ir than for RF.

The unit of infrared power intensity, or radiant intensity, is mW/sr, with sr being the abbreviation for steradian. The steradian is the unit of solid angle measure, and this is the key concept in understanding the link budget for Ir communication. As shown in Figure 5-1, the solid angle (S) subtended by an area A on the surface of a sphere of radius R is given as:

$$S = A / R^2 \quad \text{steradians} \tag{5.1}$$

$$A = 2\pi R^2 (1-\cos(a))$$

$$\text{So} \quad S = 2\pi (1-\cos(a)) \tag{5.2}$$

Note from Eq. 5.1 that at a distance of 1 metre a solid angle of 1 steradian subtends an area of 1 m^2. For small solid angles, the area A on

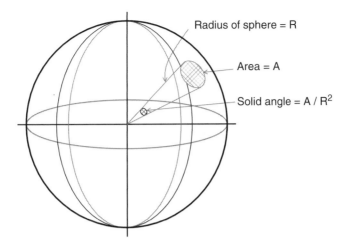

Figure 5-1: Solid Angle Subtended by an Area

the sphere can be approximated by the area of the flat circle of radius r, giving:

$$S = \pi r^2 / R^2 \text{ steradians}$$

As an example, the IrDA physical layer standard specifies a half angle (a) of between 15° and 30°. For 15°, $S = 2\pi (1-\cos(15°)) = 0.214$ steradians.

For an LED with a given emitter power density or radiant intensity, I_e, in mW/sr, the equivalent power density in mW/m² will be approximately given as:

$$P = I_e / R^2 \text{ mW/m}^2 \tag{5.3}$$

Emitter Beam Pattern
Similar to an RF antenna, an LED has a beam pattern in which radiated power drops off with increasing angle off-axis. In the example shown in Figure 5-2, the power density drops to about 85% of the on-axis value at an off-axis angle of 15°.

Inverse Square Loss
Equation 5.3 shows that the on-axis power density is inversely proportional to the square of the distance from the source. If R is doubled, the power density P drops by a factor of 4, as shown in Figure 5-3. This is the equivalent of the free space loss term in the RF link budget.

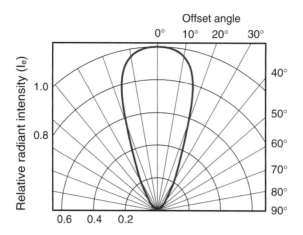

Figure 5-2: Typical LED Emitted Power Polar Diagram

Ir Detector Sensitivity

The standard detection device for high speed Ir communications is the photodiode, which has a detection sensitivity, or minimum threshold irradiance E_e, expressed in $\mu W/cm^2$. In standard power mode (see the Section "IrDA PHY Layer, p. 282"), the IrDA standard specifies a minimum emitter power of 40 mW/sr. From Eq. 5.3, the minimum power density at a receiving photodiode at a range of 1 metre will be 40 mW/m², or 4 µW/cm².

The sensitivity of a photo diode detector depends on the incident angle of the infrared source relative to the detector axis in a similar manner to that shown in Figure 5-2 for the beam pattern of an LED. Photodiode sensitivity also depends on the incident infrared wavelength as shown for example in Figure 5-4, and in any application a detector will be chosen

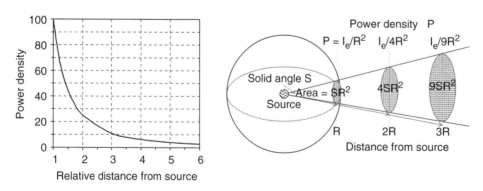

Figure 5-3: Inverse Square Distance Relationship of Radiant Power Density

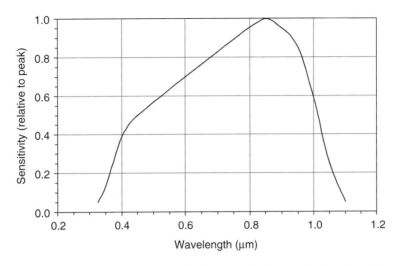

Figure 5-4: Typical Photodiode Sensitivity vs. Wavelength

with a peak spectral sensitivity close to the wavelength of the emitting device.

Ir Link Distance

The maximum link distance for an Ir link can be calculated as the distance R at which the equivalent power density (P) drops to the level of the detector's minimum threshold irradiance (E_e). Eq. 5.3 gives:

$$E_e = I_e / R^2 \text{ mW/m}^2$$

or

$$R = (I_e / E_e)^{1/2} \text{ m} \tag{5.4}$$

The effective range of an Ir link can be increased substantially, up to several tens of metres, using lenses to collimate the transmitted beam and focus the beam onto the receiving photodiode. Alignment of the lenses and of the transmitting and receiving diodes will be critical to the effectiveness of such a system. As shown in Figure 5-5, a misalignment of approximately $1/3°$ would be sufficient to break a focussed link over a range of 10 m.

Ir areal coverage may be increased in a home or small office environment by reflecting the Ir beam from a wall or ceiling in order to access a number of devices. In order to preserve power in the reflection it will be

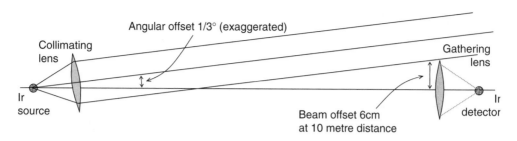

Figure 5-5: Focussed Ir Link Alignment over a 10 m Range

important to use a high reflectance, low absorption material to reflect the beam. A white painted ceiling is a good reflector of sunlight, with a reflection coefficient for visible light of around 0.9, but is a poor reflector at Ir wavelengths, absorbing about 90% of the incident Ir radiation. To achieve a comparable 0.9 reflection coefficient for Ir, an aluminium or aluminium foil covered panel would be a suitable reflector.

Summary of Part II

The radio frequency or infrared communication technologies described in Part II are at the heart of the physical layer of wireless networks. An understanding of the basics of these technologies will provide a firm foundation from which to discuss the implementation of wireless networks, whether local, personal or metropolitan area (LAN, PAN or MAN).

In particular, the link budget calculation will be important in establishing power requirements and coverage in LAN and MAN applications.

Spread spectrum and digital modulation techniques are key to understanding how a wide range of data rates can be accommodated within the limited available bandwidth, for example in the 2.4 and 5.8 GHz ISM bands.

Ultra wideband radio — using a bandwidth approaching 7 GHz with a transmitted power density below the FCC allowable noise emission level — stretches the conceptual boundaries of what radio frequency communications can achieve, and the increasing number of practical applications including wireless USB and ZigBee, are testament to the practicality of this remarkable technology.

Infrared communication links are perhaps the most "transparent" in terms of a very low requirement for user configuration, and to some extent this means that the user can be unconcerned about the underlying technology. However, even infrared links can be stretched to deliver performance over significant distances (tens of metres), given an understanding of the characteristics of Ir transmitters, detectors and infrared propagation.

WIRELESS LAN IMPLEMENTATION

Introduction

Wireless networking technology at the local area scale is perhaps the most widespread, most commercially significant and most well developed of all wireless networking technologies. In less than a decade since the ratification of the IEEE 802.11a and b standards in 1999, some 200 million 802.11 chipsets have been shipped, a business sector has grown to an estimated turnover in excess of $800 million (chipset cost alone) in 2005, and the standard has developed to the point where a 600-fold increase in data capacity, vehicular speed roaming and mesh networking are now on the horizon. Part III looks at the technologies and practical considerations that underpin successful wireless LANs.

The main technical features of wireless LAN standards are described in Chapter 6. This is an area now firmly dominated by the IEEE 802.11 standards, starting with the original 802.11b based Wi-Fi, and now including the improved security features of 802.11i and the upcoming enhancements to throughput and mobility in progress with 802.11n, r and s. Non-IEEE WLAN standards are also reviewed, the brevity of this section being a clear indication of the dominance of the IEEE standards on the local area scale.

Chapter 7 covers the implementation of WLANs, from the point of view of a medium scale corporate network, starting with the definition of user and technical requirements, through planning and installation to operation and support. This chapter concludes with a case study that looks at the specific requirements for a voice over WLAN (VoWLAN) application.

Technologies providing security for wireless LANs are described in Chapter 8, including the most recent encryption and authentication mechanisms of 802.11i. Practical WLAN security measures are then described, including checklists covering management, technical and operational security measures.

The last chapter in Part III addresses WLAN troubleshooting, and covers strategies for problem identification and diagnosis, as well as specific measures for the two most common categories of WLAN problems — connectivity and performance.

CHAPTER 6

Wireless LAN Standards

The 802.11 WLAN Standards

Origins and Evolution

The development of wireless LAN standards by the IEEE began in the late 1980s, following the opening up of the three ISM radio bands for unlicensed use by the FCC in 1985, and reached a major milestone in 1997 with the approval and publication of the 802.11 standard. This standard, which initially specified modest data rates of 1 and 2 Mbps, has been enhanced over the years, the many revisions being denoted by the addition of a suffix letter to the original 802.11, as for example in 802.11a, b and g.

The 802.11a and 802.11b extensions were ratified in July 1999, and 802.11b, offering data rates up to 11 Mbps, became the first standard with products to market under the Wi-Fi banner. The 802.11g specification was ratified in June 2003 and raised the PHY layer data rate to 54 Mbps, while offering a degree of interoperability with 802.11b equipment with which it shares the 2.4 GHz ISM band.

Table 6-1 summarises the 802.11 standard's relentless march through the alphabet, with various revisions addressing issues such as security, local regulatory compliance and mesh networking, as well as other enhancements that will lift the PHY layer data rate to 600 Mbps.

Overview of the Main Characteristics of 802.11 WLANs

The 802.11 standards cover the PHY and MAC layer definition for local area wireless networking. As shown in Figure 6-1, the upper part of the Data Link layer (OSI Layer 2) is provided by Logical Link Control (LLC)

Table 6-1: The IEEE 802.11 Standard Suite

Standard	Key features
802.11a	High speed WLAN standard, supporting 54 Mbps data rate using OFDM modulation in the 5 GHz ISM band.
802.11b	The original Wi-Fi standard, providing 11 Mbps using DSSS and CCK on the 2.4 GHz ISM band.
802.11d	Enables MAC level configuration of allowed frequencies, power levels and signal bandwidth to comply with local RF regulations, thereby facilitating international roaming.
802.11e	Addresses quality of service (QoS) requirements for all 802.11 radio interfaces, providing TDMA to prioritise and error-correction to enhance performance of delay sensitive applications.
802.11f	Defines recommended practices and an Inter-Access Point Protocol to enable access points to exchange the information required to support distribution system services. Ensures inter-operability of access points from multiple vendors, for example to support roaming.
802.11g	Enhances data rate to 54 Mbps using OFDM modulation on the 2.4 GHz ISM band. Interoperable in the same network with 802.11b equipment.
802.11h	Spectrum management in the 5 GHz band, using dynamic frequency selection (DFS) and transmit power control (TPC) to meet European requirements to minimise interference with military radar and satellite communications.
802.11i	Addresses the security weaknesses in user authentication and encryption protocols. The standard employs advanced encryption standard (AES) and 802.1x authentication.
802.11j	Japanese regulatory extension to 802.11a adding RF channels between 4.9 and 5.0 GHz.
802.11k	Specifies network performance optimisation through channel selection, roaming and TPC. Overall network throughput is maximised by efficiently loading all access points in a network, including those with weaker signal strength.
802.11n	Provides higher data rates of 150, 350 and up to 600 Mbps using MIMO radio technology, wider RF channels and protocol stack improvements, while maintaining backward compatibility with 802.11 a, b and g.
802.11p	Wireless access for the vehicular environment (WAVE), providing communication between vehicles or from a vehicle to a roadside access point using the licensed intelligent transportation systems (ITS) band at 5.9 GHz.

Table 6-1: The IEEE 802.11 Standard Suite — cont'd

802.11r	Enables fast BSS to BSS (Basic Service Set) transitions for mobile devices, to support delay sensitive services such as VoIP on stations roaming between access points.
802.11s	Extending 802.11 MAC to support ESS (Extended Service Set) mesh networking. The 802.11s protocol will enable message delivery over self-configuring multi-hop mesh topologies.
802.11T	Recommended practices on measurement methods, performance metrics and test procedures to assess the performance of 802.11 equipment and networks. The capital T denotes a recommended practice rather than a technical standard.
802.11u	Amendments to both PHY and MAC layers to provide a generic and standardised approach to inter-working with non-802.11 networks, such as Bluetooth, ZigBee and WiMAX.
802.11v	Enhancements to increase throughput, reduce interference and improve reliability through network management.
802.11w	Increased network security by extending 802.11 protection to management as well as data frames.

services specified in the 802.2 standard, which are also used by Ethernet (802.3) networks, and provide the link to the Network layer and higher layer protocols.

802.11 networks are composed of three basic components; stations, access points and a distribution system, as described in Table 6-2.

In the 802.11 standard, WLANs are based on a cellular structure where each cell, under the control of an access point, is known as a basic service

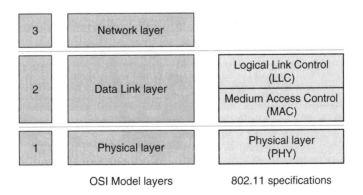

OSI Model layers   802.11 specifications

Figure 6-1: 802.11 Logical Architecture

set (BSS). When a number of stations are working in a BSS it means that they all transmit and receive on the same RF channel, use a common BSSID, use the same set of data rates and are all synchronised to a common timer. These BSS parameters are included in "beacon frames" that are broadcast at regular intervals either by individual stations or by the access point.

The standard defines two modes of operation for a BSS; ad-hoc mode and infrastructure mode. An ad-hoc network is formed when a group of two or more 802.11 stations communicate directly with each other with no access point or connection to a wired network.

This operating mode (also known as peer-to-peer mode) allows wireless connections to be quickly established for data sharing among a group of wireless enabled computers (Figure 6-2). Under ad-hoc mode the service set is called an independent basic service set (IBSS), and in an IBSS all stations broadcast beacon packets, and use a randomly generated BSSID.

Infrastructure mode exists when stations are communicating with an access point rather than directly with each other. A home WLAN with an access point and several wired devices connected through an Ethernet hub or switch is a simple example of a BSS in infrastructure mode (Figure 6-3). All communication between stations in a BSS goes through the access point, even if two wireless stations in the same cell need to communicate with each other.

This doubling-up of communication within a cell (first from sending station to the access point, then from the access point to the

Table 6-2: 802.11 Network Components

Component	*Description*
Station	Any device that implements the 802.11 MAC and PHY layer protocols.
Access point	A station that provides an addressable interface between a set of stations, known as a basic service set (BSS), and the distribution system.
Distribution system	A network component, commonly a wired Ethernet, that connects access points and their associated BSSs to form an extended service set (ESS).

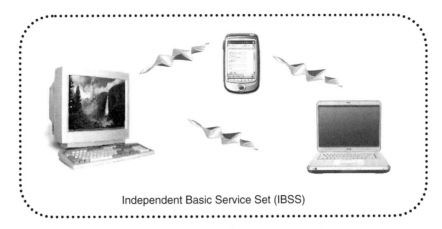

Independent Basic Service Set (IBSS)

Figure 6-2: Ad-hoc Mode Topology

destination station) might seem like an unnecessary overhead for a simple network, but among the benefits of using a BSS rather than an IBSS is that the access point can buffer data if the receiving station is in standby mode, temporarily out of range or switched off. In infrastructure mode, the access point takes on the role of broadcasting beacon frames.

The access point will also be connected to a distribution system which will usually be a wired network, but could also be a wireless bridge to other WLAN cells. In this case the cell supported by each access point is a BSS and if two or more such cells exist on a LAN the combined set is known as an extended service set (ESS).

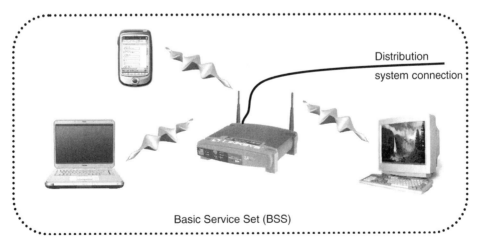

Distribution
system connection

Basic Service Set (BSS)

Figure 6-3: Infrastructure Mode Topology

In an ESS, access points (APs) will use the distribution system to transfer data from one BSS to another, and also to enable stations to move from one AP to another without any interruption in service. The transport and routing protocols that operate on the external network have no concept of mobility — of the route to a device changing rapidly — and within the 802.11 architecture the ESS provides this mobility to stations while keeping it invisible to the outside network.

Prior to 802.11k, support for mobility within 802.11 networks was limited to movement of a station between BSSs within a single ESS, so-called BSS transitions. With 802.11k, which will be described further in the Section "Network Performance and Roaming (802.11k and 802.11r, p. 162)", the roaming of stations between ESSs is supported. When a station is sensed as moving out of range, an access point is able to deliver a site report that identifies alternative access points the station can connect to for uninterrupted service.

The 802.11 MAC Layer

The MAC layer is implemented in every 802.11 station, and enables the station to establish a network or join a pre-existing network and to transmit data passed down by Logical Link Control (LLC). These functions are delivered using two classes of services, station services and distribution system services, which are implemented by the transmission of a variety of management, control and data frames between MAC layers in communicating stations.

Before these MAC services can be invoked, the MAC first needs to gain access to the wireless medium within a BSS, with potentially many other stations also competing for access to the medium. The mechanisms to efficiently share access within a BSS are described in the next section.

Wireless Media Access

Sharing media access among many transmitting stations in a wireless network is more complex to achieve than in a wired network. This is because a wireless network station is not able to detect a collision between its transmission and the transmission from another station, since a radio transceiver is unable both to transmit and to listen for other stations transmitting at the same time.

In a wired network a network interface is able to detect collisions by sensing the carrier, for example the Ethernet cable, during transmission and ceasing transmission if a collision is detected. This results in a medium access mechanism known as carrier sense multiple access/ collision detection (CSMA/CD).

The 802.11 standard defines a number of MAC layer coordination functions to co-ordinate media access among multiple stations. Media access can either be contention-based, as in the mandatory 802.11 distributed coordination function (DCF), when all stations essentially compete for access to the media, or contention free, as in the optional 802.11 point coordination function (PCF), when stations can be allocated specific periods during which they will have sole use of the media.

The media access method used by the distributed coordination function is carrier sense multiple access/collision avoidance (CSMA/CA), illustrated in Figure 6-4. In this mode a station that is waiting to transmit will sense the medium on the channel being used and wait until the medium is free of other transmissions. Once the medium is free, the station waits a predetermined period (the distributed inter-frame spacing or DIFS).

If the station senses no other transmission before the end of the DIFS period, it computes a random backoff time, between parameter values Cw_{min} and Cw_{max}, and commences its transmission if the medium remains free after this time has elapsed. The contention window parameter Cw is

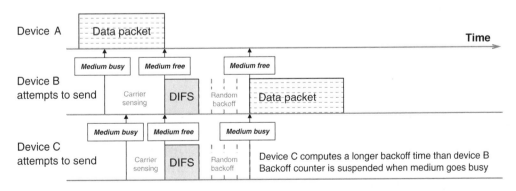

Figure 6-4: 802.11 CSMA/CA

specified in terms of a multiple of a slot time that is 20 μs for 802.11b or 9 μs for 802.11a/g networks. The backoff time is randomised so that, if many stations are waiting, they will not all try again at the same time — one will have a shorter backoff and will succeed in starting its transmission. If a station has to make repeated attempts to transmit a packet, the computed backoff period is doubled with each new attempt, up to a maximum value Cw_{max} defined for each station. This ensures that, when many stations are competing for access, individual attempts are spaced out more widely to minimise repeated collisions.

If another station is sensed transmitting before the end of the DIFS period, this is because a short IFS (SIFS) can be used by a station that is waiting either to transmit certain control frames (CTS or ACK — see Figure 6-5) or to continue the transmission of parts of a data packet that has been fragmented to improve transmission reliability.

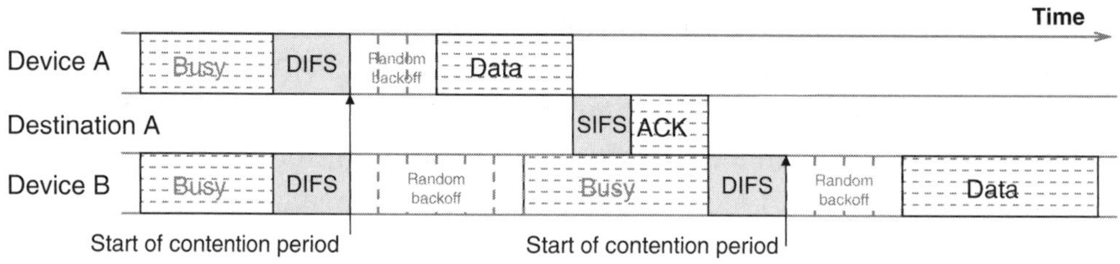

Figure 6-5: DCF Transmission Timing

CSMA/CA is a simple media access protocol that works efficiently if there is no interference and if the data being transmitted across the network is not time critical. In the presence of interference, network throughput can be dramatically reduced as stations continually backoff to avoid collisions or wait for the medium to become idle.

CSMA/CA is a contention-based protocol, since all stations have to compete for access. With the exception of the SIFS mechanism noted above, no priorities are given and, as a result, no quality of service guarantees can be made.

The 802.11 standard also specifies an optional priority based media access mechanism, the point coordination function (PCF) which is able to provide contention free media access to stations with time critical requirements. This is achieved by allowing a station implementing PCF to use an interframe spacing (PIFS) intermediate between SIFS and DIFS, effectively giving these stations higher priority access to the medium. Once the point coordinator has control, it informs all stations of the length of the contention free period, to ensure that stations do not try to take control of the medium during this period. The coordinator then sequentially polls stations, giving any pollable station the opportunity to transmit a data frame.

Although it provides some limited capability for assuring quality of service, the PCF function has not been widely implemented in 802.11 hardware and it is only with the 802.11e enhancements, described below in the Section "Quality of Service (802.11e specification), p. 157", that quality of service (QoS) and prioritised access are more comprehensively incorporated into the 802.11 standard.

Discovering and Joining a Network

The first step for a newly activated station is to determine what other stations are within range and available for association. This can be achieved by either passive or active scanning.

In passive scanning the new station listens to each channel for a predetermined period and detects beacon frames transmitted by other stations. The beacon frame will provide a time synchronisation mark and other PHY layer parameters, such as frequency hopping pattern, to allow the two stations to communicate.

If the new station has been set up with a preferred SSID name for association, it can use active scanning by transmitting a Probe frame containing this SSID and waiting for a Probe Response frame to be returned by the preferred access point. A broadcast Probe frame can also be sent, requesting all access points within range to respond with a Probe Response. This will provide the new station with a full list of access points available. The process of authentication and association can then start — either with the preferred access point or with another access point selected by the new station or by the user from the response list.

Station Services

MAC layer station services provide functions to send and receive data units passed down by the LLC and to implement authentication and security between stations, as described in Table 6-3.

Table 6-3: 802.11 MAC Layer Station Services

Service	*Description*
Authentication	This service enables a receiving station to authenticate another station prior to association. An access point can be configured for either open system or shared key authentication. Open system authentication offers minimal security and does not validate the identity of other stations — any station that attempts to authenticate will receive authentication. Shared key authentication requires both stations to have received a secret key (e.g. a passphrase) via another secure channel such as direct user input.
Deauthentication	Prior to disassociation, a station will deauthenticate from the station that it intends to stop communication with. Both deauthentication and authentication are achieved by the exchange of management frames between the MAC layers of the two communicating stations.
Privacy	This service enables data frames and shared key authentication frames to be optionally encrypted before transmission, for example using wired equivalent privacy (WEP) or Wi-Fi protected access (WPA).
MAC service data unit delivery	A MAC service data unit (MSDU) is a unit of data passed to the MAC layer by the logical link controller. The point at which the LLC accesses MAC services (at the "top" of the MAC layer) is termed the MAC service access point or SAP. This service ensures the delivery of MSDUs between these service access points. Control frames such as RTS, CTS and ACK may be used to control the flow of frames between stations, for example in 802.11b/g mixed-mode operation.

Distribution System Services

The functionality provided by MAC distribution system services is distinct from station services in that these services extend across the distribution system rather than just between sending and receiving stations at either end of the air interface. The 802.11 distribution system services are described in Table 6-4.

802.11 PHY Layer

The initial 802.11 standard, as ratified in 1997, supported three alternative PHY layers; frequency hopping and direct sequence spread

Table 6-4: 802.11 MAC Layer Distribution System Services

Service	*Description*
Association	This service enables a logical connection to be made between a station and an access point. An access point cannot receive or deliver any data until a station has associated, since association provides the distribution system with the information necessary for delivery of data.
Disassociation	A station disassociates before leaving a network, for example when a wireless link is disabled, the network interface controller is manually disconnected or its host PC is powered down.
Reassociation	The reassociation service allows a station to change the attributes (such as supported data rates) of an existing association or to change its association from one BSS to another within an extended BSS. For example, a roaming station may change its association when it senses another access point transmitting a stronger beacon frame.
Distribution	The distribution service is used by a station to send frames to another station within the same BSS, or across the distribution system to a station in another BSS.
Integration	Integration is an extension of distribution when the access point is a portal to a non-802.11 network and the MSDU has to be transmitted across this network to its destination. The integration service provides the necessary address and media specific translation so that an 802.11 MSDU can be transmitted across the new medium and successfully received by the destination device's non-802.11 MAC.

spectrum in the 2.4 GHz band as well as an infrared PHY. All three PHYs delivered data rates of 1 and 2 Mbps.

The infrared PHY specified a wavelength in the 800–900 nm range and used a diffuse mode of propagation rather than direct alignment of infrared transceivers, as is the case in IrDA for example (Section 10.5). A connection between stations would be made via passive ceiling reflection of the infrared beam, giving a range of 10–20 metres, depending on the height of the ceiling. Pulse position modulation was specified, 16-PPM and 4-PPM respectively for the 1 and 2 Mbps data rates.

Later extensions to the standard have focused on high rate DSSS (802.11b), OFDM (802.11a and g) and OFDM plus MIMO (802.11n). These PHY layers will be described in the following sections.

802.11a PHY Layer

The 802.11a amendment to the original 802.11 standard was ratified in 1999 and the first 802.11a compliant chipsets were introduced by Atheros in 2001. The 802.11a standard specifies a PHY layer based on orthogonal frequency division multiplexing (OFDM) in the 5 GHz frequency range. In the US, 802.11a OFDM uses the three unlicensed national information infrastructure bands (U-NII), with each band accommodating four non-overlapping channels, each of 20-MHz bandwidth. Maximum transmit power levels are specified by the FCC for each of these bands and, in view of the higher permitted power level, the four upper band channels are reserved for outdoor applications.

In Europe, in addition to the 8 channels between 5.150 and 5.350 GHz, 11 channels are available between 5.470 and 5.725 GHz (channels 100, 104, 108, 112, 116, 120, 124, 128, 132, 136, 140). European regulations on maximum power level and indoor versus outdoor use vary from country to country, but typically the 5.15–5.35 GHz band is reserved for indoor use with a maximum EIRP of 200 mW, while the 5.47–5.725 GHz band has an EIRP limit of 1W and is reserved for outdoor use.

Table 6-5: US FCC Specified U-NII Channels Used in the 802.11a OFDM PHY

RF Band	*Frequency Range (GHz)*	*Channel number*	*Centre frequency (GHz)*	*Maximum transmit power (mW)*
U-NII lower band	5.150–5.250	36 40 44 48	5.180 5.200 5.550 5.240	50
U-NII middle band	5.250–5.350	52 56 60 64	5.260 5.280 5.300 5.320	250
U-NII upper band	5.725–5.825	149 153 157 162	5.745 5.765 5.785 5.805	1000

As part of the global spectrum harmonisation drive following the 2003 ITU World Radio Communication Conference, the 5.470–5.725 GHz spectrum has also been available in the US since November 2003, subject to the implementation of the 802.11h spectrum management mechanisms described in the Section "Spectrum Management at 5 GHz (802.11h), p. 160".

Each of the 20 MHz wide channels accommodates 52 OFDM subcarriers, with a separation of 312.5 kHz (= 20 MHz/64) between centre frequencies. Four of the subcarriers are used as pilot tones, providing a reference to compensate for phase and frequency shifts, while the remaining 48 are used to carry data.

Four different modulation methods are specified, as shown in Table 6-6, which result in a range of PHY layer data rates from 6 Mbps up to 54 Mbps.

The coding rate indicates the error-correction overhead that is added to the input data stream and is equal to $m/(m+n)$ where n is the number of error correction bits applied to a data block of length m bits. For example, with a coding rate of 3/4 every 8 transmitted bits includes 6 bits of user data and 2 error correction bits.

The user data rate resulting from a given combination of modulation method and coding rate can be determined as follows, taking the

Table 6-6: 802.11a OFDM Modulation Methods, Coding and Data Rate

Modulation	Code bits per subcarrier	Code bits per OFDM symbol	Coding rate	Data bits per OFDM symbol	Data rate (Mbps)
BPSK	1	48	1/2	24	6
BPSK	1	48	3/4	36	9
QPSK	2	96	1/2	48	12
QPSK	2	96	3/4	72	18
16-QAM	4	192	1/2	96	24
16-QAM	4	192	3/4	144	36
64-QAM	6	288	2/3	192	48
64-QAM	6	288	3/4	216	54

64-QAM, 3/4 coding rate line as an example. During one symbol period of 4 µS, which includes a guard interval of 800 nS between symbols, each carrier is encoded with a phase and amplitude represented by one point on the 64-QAM constellation. Since there are 64 such points, this encodes 6 code bits. The 48 subcarriers together therefore carry $6 \times 48 = 288$ code bits for each symbol period. With a 3/4 coding rate, 216 of those code bits will be user data while the remaining 72 will be error correction bits. Transmitting 216 data bits every 4 µS corresponds to a data rate of 216 data bits per OFDM symbol $\times$ 250 OFDM symbols per second = 54 Mbps.

The 802.11a specifies 6, 12 and 24 Mbps data rates as mandatory, corresponding to 1/2 coding rate for BPSK, QPSK and 16-QAM modulation methods. The 802.11a MAC protocol allows stations to negotiate modulation parameters in order to achieve the maximum robust data rate.

Transmitting at 5 GHz gives 802.11a the advantage of less interference compared to 802.11b, operating in the more crowded 2.4 GHz ISM band, but the higher carrier frequency is not without disadvantages. It restricts 802.11a to near line-of-sight applications and, taken together with the lower penetration at 5 GHz, means that indoors more WLAN access points are likely to be required to cover a given operating area.

802.11b PHY Layer

The original 802.11 DSSS PHY used the 11-chip Barker spreading code (as described in the Section "Chipping, Spreading and Correcting, p. 80") together with DBPSK and DQPSK modulation methods to deliver PHY layer data rates of 1 and 2 Mbps respectively (Table 6-7).

The high rate DSSS PHY specified in 802.11b added complementary code keying (CCK), using 8-chip spreading codes, as described in the Section "Complementary Code Keying, p. 82".

The 802.11 standard supports dynamic rate shifting (DRS) or adaptive rate selection (ARS), allowing the data rate to be dynamically adjusted to compensate for interference or varying path losses. When interference is present, or if a station moves beyond the optimal range for reliable operation at the maximum data rate, access points will progressively fall

Table 6-7: 802.11b DSSS Modulation Methods, Coding and Data Rate

Modulation	Code length (Chips)	Code type	Symbol rate (Msps)	Data bits per symbol	Data rate (Mbps)
BPSK	11	Barker	1	1	1
QPSK	11	Barker	1	2	2
DQPSK	8	CCK	1.375	4	5.5
DQPSK	8	CCK	1.375	8	11

back to lower rates until reliable communication is restored. This strategy is based on the implications of Eq. (4-1), which showed that *SNR* is proportional to the transmitted energy per bit, so that by falling back to a lower data rate, a higher *SNR* and lower *BER* can be achieved.

Conversely, if a station moves back within range for a higher rate, or if interference is reduced, the link will shift to a higher rate. Rate shifting is implemented in the PHY layer and is transparent to the upper layers of the protocol stack.

The 802.11 standard specifies the division of the 2.4 GHz ISM band into a number of overlapping 22 MHz channels, as shown in Figure 4-9. The FCC in the US and the ETSI in Europe have both authorised the use of spectrum from 2.400 to 2.4835 GHz, with 11 channels approved in the US and 13 in (most of) Europe. In Japan, channel 14 at 2.484 GHz is also authorised by the ARIB. Some countries in Europe have more restrictive channel allocations, notably France where only four channels (10 through 13) are approved. The available channels for 802.11b operation are summarised in Table 6-8.

The 802.11b standard also includes a second, optional modulation and coding method, packet binary convolutional coding (PBCC™–Texas Instruments), which offers improved performance at 5.5 and 11 Mbps by achieving an additional 3 dB processing gain. Rather than the 2 or 4 phase states or phase shifts used by BPSK/DQSK, PBCC uses 8-PSK (8 phase states) giving a higher chip per symbol rate. This can be translated into either a higher data rate for a given chipping code length, or a higher processing gain for a given data rate, by using a longer chipping code.

**Table 6-8: International Channel Availability for 802.11b
Networks in the 2.4 GHz Band**

Channel number	Centre frequency (GHz)	Geographical usage
1	2.412	US, Canada, Europe, Japan
2	2.417	US, Canada, Europe, Japan
3	2.422	US, Canada, Europe, Japan
4	2.427	US, Canada, Europe, Japan
5	2.432	US, Canada, Europe, Japan
6	2.437	US, Canada, Europe, Japan
7	2.442	US, Canada, Europe, Japan
8	2.447	US, Canada, Europe, Japan
9	2.452	US, Canada, Europe, Japan
10	2.457	US, Canada, Europe, Japan, France
11	2.462	US, Canada, Europe, Japan, France
12	2.467	Europe, Japan, France
13	2.472	Europe, Japan, France
14	2.484	Japan

802.11g PHY Layer

The 802.11g PHY layer was the third 802.11 standard to be approved by the IEEE standards board and was ratified in June 2003. Like 802.11b, 11g operates in the 2.4 GHz band, but increases the PHY layer data rate to 54 Mbps, as for 802.11a.

The 802.11g uses OFDM to add data rates from 12 Mbps to 54 Mbps, but is fully backward compatible with 802.11b, so that hardware supporting both standards can operate in the same 2.4 GHz WLAN. The OFDM modulation and coding scheme is identical to that applied in the 802.11a standard, with each 20 MHz channel in the 2.4 GHz band (as shown in Table 6-8) divided into 52 subcarriers, with 4 pilot tones and 48 data tones. Data rates from 6 to 54 Mbps are achieved using the same modulation methods and coding rates shown for 802.11a in Table 6-6.

Although 802.11b and 11g hardware can operate in the same WLAN, throughput is reduced when 802.11b stations are associated with an 11g network (so-called mixed-mode operation) because of a number of protection mechanisms to ensure interoperability, as described Table 6-9.

Table 6-9: 802.11b/g Mixed-Mode Interoperability Mechanisms

Mechanism	*Description*
RTS/CTS	Before transmitting, 11b stations request access to the medium by sending a request to send (RTS) message to the access point. Transmission can commence on receipt of the clear to send (CTS) response. This avoids collisions between 11b and 11g transmissions, but the additional RTS/CTS signalling adds a significant overhead that decreases network throughput.
CTS to self	The CTS to self option dispenses with the exchange of RTS/CTS messages and just relies on the 802.11b station to check that the channel is clear before transmitting. Although this does not provide the same degree of collision avoidance, it can increase throughput significantly when there are fewer stations competing for medium access.
Backoff time	802.11g backoff timing is based on the 802.11a specification (up to a maximum of 15×9 µS slots) but in mixed-mode an 802.11g network will adopt 802.11b backoff parameters (maximum 31×20 µS slots). The longer 802.11b backoff results in reduced network throughput.

The impact of mixed mode operation on the throughput of an 802.11g network is shown in Table 6-10.

A number of hardware manufacturers have introduced proprietary extensions to the 802.11g specification to boost the data rate above 54 Mbps. An example is D-Link's proprietary "108G" which uses packet bursting and channel bonding to achieve a PHY layer data rate of 108 Mbps. Packet bursting, also known as frame bursting, bundles short data packets into fewer but larger packet to reduce the impact of gaps between transmitted packets.

Packet bursting as a data rate enhancement strategy runs counter to packet fragmentation as a strategy for improving transmission robustness, so packet bursting will only be effective when interference or high levels of contention between stations are absent.

Channel bonding is a method where multiple network interfaces in a single machine are used together to transmit a single data stream. In the 108G example, two non-overlapping channels in the 2.4 GHz ISM band are used simultaneously to transmit data frames.

Data Rates at the PHY and MAC Layer

In considering the technical requirements for a WLAN implementation in Chapter 7, it will be important to recognise the difference between the headline data rate of a wireless networking standard and the true effective data rate as seen by the higher OSI layers when passing data packets down to the MAC layer.

Each "raw" data packet passed to the MAC service access point (MAC SAP) will acquire a MAC header and a message integrity code and additional security related header information before being passed to the PHY layer for transmission. The headline data rate, for example 54 Mbps for 802.11a or 11g networks, measures the transmission rate of this extended data stream at the PHY layer.

The effective data rate is the rate at which the underlying user data is being transmitted if all the transmitted bits relating to headers, integrity checking and other overheads are ignored. For example, on average, every 6 bits of raw data passed to the MAC SAP of an 802.11b WLAN will gain an extra 5 bits of overhead before transmission, reducing a PHY layer peak data rate of 11 Mbps to an effective rate of 6 Mbps.

Table 6-10 shows the PHY and MAC SAP data rates for 802.11 WLANs. For 802.11g networks the MAC SAP data rate depends on the presence of 802.11b stations, as a result of the mixed-mode media access control mechanisms described in the previous section.

802.11 Enhancements

In the following sections some of the key enhancements to 802.11 network capabilities and performance will be described. The security enhancements covered by the 802.11i update are described separately in Chapter 8, which is devoted to WLAN security.

Table 6-10: PHY and MAC SAP Throughput Comparison for 802.11a, b and g Networks

Network standard and configuration	PHY data rate (Mbps)	Effective MAC SAP throughput (Mbps)	Effective Throughput versus 802.11b (%)
802.11b network	11	6	100
802.11g network with 802.11b stations (CTS/RTS)	54	8	133
802.11g network with 802.11b stations (CTS-to-self)	54	13	217
802.11g network with no 802.11b stations	54	22	367
802.11a	54	25	417

Quality of Service (802.11e specification)

The 802.11e specification provides a number of enhancements to the 802.11 MAC to improve the quality of service for time sensitive applications, such as streaming media and voice over wireless IP (VoWIP), and was approved for publication by the IEEE Standards Board in September 2005.

The 802.11e specification defines two new coordination functions for controlling and prioritising media access, which enhance the original 802.11 DCF and PCF mechanisms described above in Section "Wireless Media Access, p. 144". Up to eight traffic classes (TC) or access categories (AC) are defined, each of which can have specific QoS requirements and receive specific priority for media access.

The simplest of the 802.11e coordination functions is enhanced DCF (EDCF) which allows several MAC parameters determining ease of media access to be specified per traffic class. An arbitrary interframe space (AIFS) is defined which is equal to DIFS for the highest priority traffic class and longer for other classes. This provides a deterministic mechanism for traffic prioritisation as shown in Figure 6-6.

The minimum backoff time Cw_{min} is also TC dependent, so that, when a collision occurs, higher priority traffic, with a lower Cw_{min}, will have a higher probability of accessing the medium.

Time

Device A	⌐ Busy ¬		
		SIFS	Interframe spacing for control frames and continuation of high priority transmission
Device B		PIFS	Interframe spacing for Point Coordination Function
Device C		DIFS	High priority traffic; DIFS and short contention period
Device D		AIFS[2]	Medium priority traffic; AIFS[2] and intermediate contention period
Device E		AIFS[3]	Low priority traffic; AIFS[3] and longer contention period

Figure 6-6: EDCF Timing

Each station maintains a separate queue for each TC (Figure 6-7), and these behave as virtual stations, each with their individual MAC parameters. If two queues within a station reach the end of their backoff periods at the same time, data from the higher priority queue will be transmitted when the station gains access to the medium.

Although the EDCF coordination mode does not provide a guaranteed service for any TC, it has the advantage of being simple to configure and implement as an extension of DCF.

The second enhancement defines a new hybrid coordination function (HCF) which complements the polling concept of PCF with an awareness of the

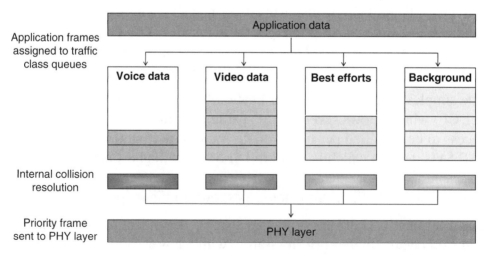

Figure 6-7: EDCF Traffic Class Queues

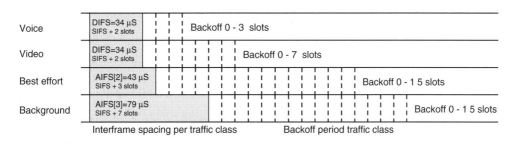

Figure 6-8: AIFS and Backoff Timing per WMM Traffic Class

QoS requirements of each station. Stations report the lengths of their queues for each traffic class and the hybrid coordinator uses this to determine which station will receive a transmit opportunity (TXOP) during a contention free transmission period. This HCF controlled channel access (HCCA) mechanism considers several factors in determining this allocation;

- The priority of the TC

- The QoS requirements of the TC (bandwidth, latency and jitter)

- Queue lengths per TC and per station

- The duration of the TXOP available to be allocated

- The past QoS given to the TC and station.

HCCA allows applications to schedule access according to their needs, and therefore enables QoS guarantees to be made. Scheduled access requires a client station to know its resource requirements in advance and scheduling concurrent traffic from multiple stations also requires the access point to make certain assumptions regarding data packet sizes, data transmission rates and the need to reserve surplus bandwidth for transmission retries.

The Wi-Fi Alliance adopted a subset of the 802.11e standard in advance of the IEEE's September 2005 approval. This subset, called Wi-Fi multimedia (WMM), describes four access categories as shown in Table 6-11, with EDCF timings as shown in Figure 6-8.

The prioritisation mechanism certified in WMM is equivalent to the EDCF coordination mode defined in 802.11e but did not initially include the scheduled access capability available through HCF and HCCA. This and other 802.11e capabilities are planned to be progressively included in the Wi-Fi Alliance's WMM certification program.

Table 6-11: WMM Access Category Descriptions

Access category	*Description*
WMM voice priority	Highest priority. Allows multiple concurrent VoWLAN calls, with low latency and quality equal to a toll voice call.
WMM video priority	Prioritises video traffic above lower categories. One 802.11g or 802.11a channel can support 3 to 4 standard definition TV streams or 1 high definition TV stream.
WMM best effort priority	Traffic from legacy devices, from applications or devices that lack QoS capabilities, or traffic such as internet surfing that is less sensitive to latency but is affected by long delays.
WMM background priority	Low priority traffic, such as a file download or print job, that does not have strict latency or throughput requirements.

Spectrum Management at 5 GHz (802.11h)

The 802.11h standard supplements the 802.11 MAC with two additional spectrum management services, transmit power control (TPC), which limits the transmitted power to the lowest level needed to ensure adequate signal strength at the farthest station, and dynamic frequency selection (DFS), which enables a station to switch to a new channel if it detects other non-associated stations or systems transmitting on the same channel.

These mechanisms are required for 5 GHz WLANs operating under European regulations, in order to minimise interference with satellite communications (TPC) and military radar (DFS), and support for the 802.11h extensions was required from 2005 for all 802.11a compliant systems operating in Europe.

In the US, compliance with 802.11h is also required for 802.11a products operating in the 12 channels from 5.47 to 5.725 GHz. IEEE 802.11h compliant networks therefore have access to 24 non-overlapping OFDM channels, resulting in a potential doubling of overall network capacity.

Transmit Power Control

An 802.11h compliant station indicates its transmit power capability, including minimum and maximum transmit power levels in dBm, in the association or reassociation frame sent to an access point. An access point may refuse the association request if the station's transmit power capability is unacceptable, for example if it violates local constraints. The access point in return indicates local maximum transmit power constraints in its beacon and probe request frames.

An access point monitors signal strength within its BSS by requesting stations to report back the link margin for the frame containing the report request and the transmit power used to transmit the report frame back to the access point. This data is used by the access point to estimate the path loss to other stations and to dynamically adjust transmit power levels in its BSS in order to reduce interference with other devices while maintaining sufficient link margin for robust communication.

Dynamic Frequency Selection

When a station uses a Probe frame to identify access points in range, an access point will specify in the Probe Response frame that it uses DFS. When a station associates or re-associates with an access point that uses DFS, the station provides a list of supported channels that enables the access point to determine the best channel when a shift is required. As for TPC, an access point may reject an association request if a station's list of supported channels is considered unacceptable, for example if it is too limited.

To determine if other radio transmissions are present, either on the channel in use or on a potential new channel, an access point sends a measurement request to one or more stations identifying the channel where activity is to be measured, and the start time and duration of the measurement period. To enable these measurements, the access point can specify a quiet period in its beacon frames to ensure that all other associated stations stop transmission during the measurement period. After performing the requested measurement, stations send back a report on the measured channel activity to the access point.

When necessary, channel switching is initiated by the access point, which sends a channel switch announcement management frame to all associated stations. This announcement identifies the new channel, specifies the number of beacon periods until the channel switch takes effect, and also specifies whether or not further transmissions are allowed on the current channel. The access point can use the short interframe spacing (SIFS — see the Section "Wireless Media Access, p. 144") to gain priority access to the wireless medium in order to broadcast a channel switch announcement.

Dynamic frequency selection is more complicated in an IBSS (ad-hoc mode) as there is no association process during which supported channel information can be exchanged, and no access point to coordinate channel measurement or switching. A separate DFS owner service is defined in 802.11h to address these complications, although channel switching remains inherently less robust in an IBSS than in an infrastructure mode BSS.

Network Performance and Roaming (802.11k and 802.11r)
A client station may need to make a transition between WLAN access points for one of three reasons, as described in Table 6-12.

The 802.11 Task Groups TGk and TGr are addressing issues relating to handoffs or transitions between access points that need to be fast and reliable for applications such as VoWLAN. TGk will standardise radio measurements and reports that will enable location-based services, such as a roaming station's choice of a new access point to connect to, while TGr aims to minimise the delay and maintain QoS during these transitions.

802.11k; Radio Resource Measurement Enhancements
The 802.11 Task Group TGk, subtitled Radio Resource Measurement Enhancements, began meeting in early 2003 with the objective of

Table 6-12: Reasons for Roaming in a WLAN

Roaming need	*Description*
Mobile client station	A mobile client station may move out of range of its current access point and need to transition to another access point with a higher signal strength.
Service availability	The QoS available at the current access point may either deteriorate or may be inadequate for a new service requirement, for example if a VoWLAN application is started.
Load balancing	An access point may redirect some associated clients to another available access point in order to maximize the use of available capacity within the network.

defining radio and network information gathering and reporting mechanisms to aid the management and maintenance of wireless LANs.

The 11k supplement will be compatible with the 802.11 MAC as well as implementing all mandatory parts of the 802.11 standards and specifications, and targets improved network management for all 802.11 networks. The key measurements and reports defined by the supplement are as follows;

- Beacon reports
- Channel reports
- Hidden station reports
- Client station statistics
- Site reports.

IEEE 802.11k will also extend the 802.11h TPC to cover other regulatory requirements and frequency bands.

Stations will be able to use these reports and statistics to make intelligent roaming decisions, for example eliminating a candidate access point if a high level of non-802.11 energy is detected in the channel being used. The 802.11k supplement only addresses the measurement and reporting of this information and does not address the processes and decisions that will make use of the measurements.

The three roaming scenarios described above will be enabled by the TGk measurements and reports, summarised in Table 6-13.

For example, a mobile station experiencing a reduced RSSI will request a neighbour report from its current access point that will provide information on other access points in its vicinity. A smart roaming algorithm in the mobile station will then analyse channel conditions and the loading of candidate access points and select a new access point that is best able to provide the required QoS.

Once a new access point has been selected, the station will perform a BSS transition by disassociating from the current access point and associating with the new one, including authentication and establishing the required QoS.

Table 6-13: 802.11k Measurements and Reports

802.11k feature	*Description*
Beacon report	Access points will use a beacon request to ask a station to report all the access point beacons it detects on a specified channel. Details such as supported services, encryption types and received signal strength will be gathered.
Channel reports (noise histogram, medium sensing time histogram report and channel load report)	Access points can request stations to construct a noise histogram showing all non-802.11 energy detected on a specified channel, or to report data about channel loading (how long a channel was busy during a specified time interval as well as the histogram of channel busy and idle times).
Hidden station report	Under 802.11k, stations will maintain lists of hidden stations (stations that they can detect but are not detected by their access point). Access points can request a station to report this list and can use the information as input to roaming decisions.
Station statistic report and frame report	802.11k access points will be able to query stations to report statistics such as the link quality and network performance experienced by a station, the counts of packets transmitted and received, and transmission retries.
Site report	A station can request an access point to provide a site report — a ranked list of alternative access points based on an analysis of all the data and measurements available via the above reports.

802.11r; Fast BSS Transitions

The speed and security of transitions between access points will be further enhanced by the 802.11r specification which is also under development and is intended to improve WLAN support for mobile telephony via VoWLAN. IEEE 802.11r will give access points and stations the ability to make fast BSS to BSS transitions through a four-step process;

- Active or passive scanning for other access points in the vicinity,

- Authentication with one or more target access points,

- Reassociation to establish a connection with the target access point, and

- Pairwise temporal key (PTK) derivation and 802.1x based authentication via a 4-way handshake, leading to re-establishment of the connection with continuous QoS through the transition.

A key element of the process of associating with the new access point will be a pre-allocation of media reservations that will assure continuity of service — a station will not be in the position of having jumped to a new access point only to find it is unable to get the slot time required to maintain a time critical service.

The 802.11k and 802.11r enhancements address roaming within 802.11 networks, and are a step towards transparent roaming between different wireless networks such as 802.11, 3G and WiMAX. The IEEE 802.21 media independent handover (MIH) function, which is described further in the Section "Network Independent Roaming, p. 347", will eventually enable mobile stations to roam across these diverse wireless networks.

MIMO and Data Rates to 600 Mbps (802.11n)
The IEEE 802.11 Task Group TGn started work during the second half of 2003 to respond to the demand for further increase in WLAN performance, and aims to deliver a minimum effective data rate of 100 Mbps through modifications to the 802.11 PHY and MAC layers.

This target data rate, at the MAC service entry point (MAC SAP), will require a PHY layer data rate in excess of 200 Mbps, representing a fourfold increase in throughput compared to 802.11a and 11g networks. Backward compatibility with 11a/b/g networks will ensure a smooth transition from legacy systems, without imposing excessive performance penalties on the high rate capable parts of a network.

Although there is still considerable debate among the supporters of alternative proposals, the main industry group working to accelerate the development of the 802.11n standard is the enhanced wireless consortium (EWC) which published Rev 1 of its MAC and PHY proposals in September 2005. The following description is based on the EWC proposals.

The two key technologies that are expected to be required to deliver the aspired 802.11n data rate are multi-input multi-output (MIMO) radio and OFDM with extended channel bandwidths.

MIMO radio, outlined in the Section "MIMO Radio, p. 124" is able to resolve information transmitted over several signal paths using multiple spatially separated transmitter and receiver antennas. The use of multiple antennas provides an additional gain (the diversity gain) that increases the receiver's ability to decode the data stream.

The extension of channel bandwidths, most likely by the combination of two 20 MHz channels in either the 2.4 GHz or 5 GHz bands, will further increase capacity since the number of available OFDM data tones can be doubled.

To achieve a 100 Mbps effective data rate at the MAC SAP it is expected to require either a 2 transmitter × 2 receiver antenna system operating over a 40 MHz bandwidth or a 4 × 4 antenna system operating over 20 MHz, with respectively 2 or 4 spatially separated data streams being processed. In view of the significant increase in hardware and signal processing complexity in going from 2 to 4 data streams, the 40 MHz bandwidth solution is likely to be preferred where permitted by local spectrum regulations. To ensure backward compatibility, a PHY operating mode will be specified in which 802.11a/g OFDM is used in either the upper or lower 20 MHz of a 40 MHz channel.

Maximising data throughput in 802.11n networks will require intelligent mechanisms to continuously adapt parameters such as channel bandwidth and selection, antenna configuration, modulation scheme and coding rate, to varying wireless channel conditions.

A total of 32 modulation and coding schemes are initially specified, in four groups of eight, depending on whether one to four spatial streams are used. Table 6-14 shows the modulation and coding schemes for the highest rate case — four spatial streams operating over 40 MHz bandwidth providing 108 OFDM data tones. For fewer spatial streams, the data rates are simply proportional to the number of streams.

As for 802.11a/g, these data rates are achieved with a symbol period of 4.0 µS. A further data rate increase of 10/9 (e.g. from 540 to 600 Mbps) is achieved in an optional short guard interval mode, which reduces the symbol period to 3.6 µS by halving the inter-symbol guard interval from 800 nS to 400 nS.

Table 6-14: 802.11n OFDM Modulation Methods, Coding and Data Rate

Modulation	Code bits per subcarrier (per stream)	Code bits per symbol (all streams)	Coding rate	Data bits per symbol (all streams)	Data rate (Mbps)
BPSK	1	432	1/2	216	54
QPSK	2	864	1/2	432	108
QPSK	2	864	3/4	648	162
16-QAM	4	1728	1/2	864	216
16-QAM	4	1728	3/4	1296	324
64-QAM	4	2592	2/3	1728	432
64-QAM	6	2592	3/4	1944	486
64-QAM	6	2592	5/6	2160	540

MAC framing and acknowledgement overheads will also need to be reduced in order to increase MAC efficiency (defined as the effective data rate at the MAC SAP as a fraction of the PHY layer data rate). With the current MAC overhead, a PHY layer data rate approaching 500 Mbps would be required to deliver the target 100 Mbps data rate at the MAC SAP.

Mesh Networking (802.11s)
As described in the Section "The 802.11 WLAN Standards, p. 139", the 802.11 topology relies on a distribution system (DS) to link BSSs together to form an ESS. The DS is commonly a wired Ethernet linking access points (Figure 6-3), but the 802.11 standard also provides for a wireless distribution system between separated Ethernet segments by defining a four-address frame format that contains source and destination station addresses as well as the addresses of the two access points that these stations are connected to, as shown in Figure 6-9.

The objective of the 802.11s Task Group, which began working in 2004, is to extend the 802.11 MAC as the basis of a protocol to establish a wireless distribution system (WDS) that will operate over self-configuring multi-hop wireless topologies, in other words an ESS mesh.

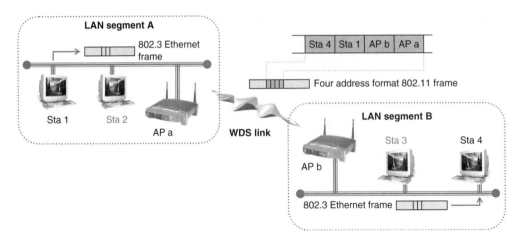

Figure 6-9: Wireless Distribution System Based on Four-Address Format MAC Frame

An ESS mesh is a collection of access points, connected by a WDS, that automatically learns about the changing topology and dynamically re-configures routing paths as stations and access points join, leave and move within the mesh. From the point of view of an individual station and its relationship with a BSS and ESS, an ESS mesh is functionally equivalent to a wired ESS.

Two industry alliances emerged during 2005 to promote alternative technical proposals for consideration by TGs; the Wi-Mesh Alliance and SEEMesh (for Simple, Efficient and Extensible Mesh).

The main elements of the Wi-Mesh proposal are a mesh coordination function (MCF) and a distributed reservation channel access protocol (DRCA) to operate alongside the HCCA and EDCA protocols (Figure 6-10). Some of the key features of the proposed Wi-Mesh MCF are summarised in Table 6-15.

The final ESS mesh specification is likely to include prioritised traffic handling based on 802.11e QoS mechanisms as well as security features and enhancements to the 802.11i standard.

The evolution of 802.11 security will be fully described in Chapter 8, but mesh networking introduces some security considerations in addition to those that have been progressively solved for non-mesh WLANs by

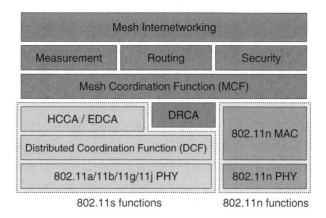

Figure 6-10: Wi-Mesh Logical Architecture

WEP, WPA, WPA2 and 802.11i. In a mesh network additional security methods are needed to identify nodes that are authorised to perform routing functions, in order to ensure a secure link for routing information messages. This will be more complicated to achieve in a mesh, where there will commonly be no centralised authentication server.

The work of the 802.11s TG is at an early stage, and ratification of the final accepted proposal is not expected before 2008.

Table 6-15: Wi-MESH Mesh Coordination Function (MCF) Features

Wi-Mesh MCF feature	*Description*
Media access coordination across multiple nodes	Media access coordination in a multi-hop network to avoid performance degradation and meet QoS guarantees.
Support for QoS	Traffic prioritisation within the mesh; flow control over multi-hop paths; load control and contention resolution mechanisms.
Efficient RF frequency and spatial reuse	To mitigate performance loss resulting from hidden and exposed stations, and allow for concurrent transmissions to enhance capacity.
Scalability	Enabling different network sizes, topologies and usage models.
PHY independent	Independent of the number of radios, channel quality, propagation environment and antenna arrangement (including smart antennas).

Other WLAN Standards

Although the wireless LAN landscape is now comprehensively dominated by the 802.11 family of standards, there was a brief period in the evolution of WLAN standards when that dominance was far from assured. From 1998 to 2000, equipment based on alternative standards briefly held sway. This short reign was brought to an end by the rapid market penetration of 802.11b products, with 10 million 802.11b based chipsets being shipped between 1999 and end-2001. The HomeRF and HiperLAN standards, which are now of mainly historical interest, are briefly described in the following sections.

HomeRF

The Home Radio Frequency (HomeRF) Working Group was formed in 1998 by a group of PC, consumer electronics and software companies, including Compaq, HP, IBM, Intel, Microsoft and Motorola, with the aim of developing a wireless network for the home networking market. The Working Group developed the specification for SWAP — Shared Wireless Access Protocol — which provided wireless voice and data networking.

SWAP was derived from the IEEE's 802.11 and ETSI's DECT (digitally enhanced cordless telephony) standards and includes MAC and PHY layer specifications with the main characteristics summarised in Table 6-16.

Table 6-16: Main Characteristics of the HomeRF SWAP

SWAP specification	*Main characteristics*
MAC	TDMA for synchronous data traffic — up to 6 TDD voice conversations. CSMA/CA for asynchronous data traffic, with prioritisation for streaming data. CSMA/CA and TDMA periods in a single SWAP frame.
PHY	FHSS radio in the 2.4 GHz ISM band. 50–100 hops per second. 2- and 4-FSK modulation deliver PHY layer data rates of 0.8 and 1.6 Mbps.

Although the HomeRF Working Group claimed some early market penetration of SWAP based products, by 2001, as SWAP 2.0 was being introduced with a 10 Mbps PHY layer data rate, the home networking market had been virtually monopolised by 802.11b products. The Working Group was finally disbanded in January 2003.

HiperLAN/2

HiperLAN stands for high performance radio local area network and is a wireless LAN standard that was developed by the European Telecommunications Standards Institute's Broadband Radio Access Networks (BRAN) project. The HiperLAN/2 Global Forum was formed in September 1999 by Bosch, Ericsson, Nokia and others, as an open industry forum to promote HiperLAN/2 and ensure completion of the standard.

The HiperLAN/2 PHY layer is very similar to the 802.11a PHY, using OFDM in the 5 GHz band to deliver a PHY layer data rate of up to 54 Mbps. The key difference between 802.11a and HiperLAN/2 is at the MAC layer where, instead of using CSMA/CA to control media access, HiperLAN/2 uses time division multiple access (TDMA). Aspects of these two access methods are compared in Table 6-17.

The technical advantages of HiperLAN/2, namely QoS, European compatibility and higher MAC SAP data rate, have now to a large

Table 6-17: CSMA/CA and TDMA Media Access Compared

Media access method	*Characteristics*
CSMA/CA	Contention based access, collisions or interference result in indefinite backoff. QoS to support synchronous (voice and video) traffic only introduced with 802.11e. MAC efficiency reduced (54 Mbps at PHY = ca. 25 Mbps at MAC SAP).
TDMA	Dynamically assigned time slot based on a station's throughput. Support for synchronous traffic. Higher MAC efficiency (54 Mbps at PHY = ca. 40 Mbps at MAC SAP). Ability to interface with 3G as well as IP networks.

Table 6-18: 802.11a, b and g Mandatory and Optional Modulation and Coding Schemes

Rate (Mbps)	802.11a		802.11b		802.11g	
	Mandatory	*Optional*	*Mandatory*	*Optional*	*Mandatory*	*Optional*
1 & 2			Barker		Barker	
5.5 & 11			CCK	PBCC	CCK	PBCC
6, 12 & 24	OFDM				OFDM	CCK-OFDM
9 & 18		OFDM				OFDM, CCK-OFDM
22 & 33						PBCC
36, 48 & 54		OFDM				OFDM, CCK-OFDM

extent been superseded by 802.11 updates, such as the QoS enhancements introduced in 802.11e (Section "Quality of Service (802.11e), p. 157") and the 802.11h PHY layer enhancements specifically introduced to cater for European regulatory requirements (Section "Spectrum Management at 5 GHz (802.11h), p. 160"). As a result, the previous support for HiperLAN/2 in the European industry has virtually disappeared.

Given the overwhelming industry focus on products based on the 802.11 suite of standards, it seems unlikely that HiperLAN/2 will ever establish a foothold in the wireless LAN market, the clearest indication of this being perhaps that Google News returns zero hits for HiperLAN/2!

Summary

Since the ratification of 802.11b in July 1999, the 802.11 standard has established a dominant and seemingly unassailable position as the basis of WLAN technology. The various 802.11 specifications draw on a wide range of applicable techniques, such as the modulation and coding schemes shown in Table 6-18, and continue to motivate the further development and deployment of new technologies, such as MIMO radio and the coordination and control functions required for mesh networking.

As future 802.11 Task Groups make a second pass through the alphabet, the further enhancement of WLAN capabilities will no doubt continue to present a rich and fascinating tapestry of technical developments.

Chapter 6 has provided a grounding in the technical aspects and capabilities of current wireless LAN technologies. Chapter 7 now builds on that foundation in describing the practical considerations in implementing wireless LANs.

CHAPTER 7

Implementing Wireless LANs

There are many routes that lead from the identification of a user requirement for wireless networking to the operation and support of an installed WLAN, and the best approach to be taken will depend on the nature and scale of the project. In this chapter a five-step process is described that is scalable from a simple ad-hoc home network to a large scale corporate WLAN, linking multiple buildings.

In small scale projects, such as in implementing a typical home or small office WLAN, some of these steps will be very short or may be eliminated altogether. Nevertheless, an awareness of the issues addressed in these steps will contribute to the successful implementation of even the smallest project.

The five key steps in the planning and implementation of a wireless LAN area are as follows;

1. Evaluating requirements and choosing the right technology

   ■ Establish the user requirements; what is it that the users want to be able to achieve and what are their expectations of performance?

   ■ Establish the technical requirements; what attributes does the technological solution need to possess in order to deliver these user requirements?

   ■ Evaluate the available technologies; how do each of the available or emerging wireless LAN technologies rank against the technical requirements?

- Selecting network hardware components; should a single or multi-vendor strategy be followed? What are the advantages and disadvantages?

2. Planning and designing the wireless LAN

- Surveying the RF environment; what other sources of RF energy or potential barriers to RF propagation are present in the target area of the wireless LAN?

- Designing the physical architecture; which architecture is right for the specific setting of the network?

3. Pilot testing

- Testing the chosen technology and architecture; does the chosen solution deliver the expected performance?

4. Implementation and configuration

- Putting the final wireless LAN in place and introducing it to the user group

- Configuring the appropriate security measures

5. Operation and support

- Keeping the wireless LAN operating efficiently and providing user support

The following sections progress through each of these planning and implementation steps, while Chapter 8 is devoted to a more detailed description of the security aspects of WLANs.

Evaluating Wireless LAN Requirements

Establishing User Requirements
If the wireless LAN is being implemented to support a large user group it will be important to gather a wide range of views on user requirements, perhaps by using a questionnaire or by interview. As a first step it may be necessary to raise awareness by demonstrating the technology to the prospective user group, so that they are better able to give an informed view on requirements.

User requirements should be expressed in terms of the user experience rather than any particular solution or technical attribute, as they are independent of specific technologies. For example, in relation to performance expectations, a PHY layer data rate is a technical attribute, whereas the transfer time for a specified large file size is what the user is really concerned about.

Common categories of user requirement are listed and discussed in Table 7-1.

In virtually all aspects of user requirements the question of future proofing also arises; are future developments expected in the users' work processes, the type of technologies deployed in the users' business, growth in the business, etc., that will change the overall demands placed on the WLAN?

Table 7-1: WLAN User Requirement Types

Requirement type	*Considerations*
Usage model	What user activities does the WLAN have to support? Are users routinely transferring large files over the network, such as Internet downloads or video editing? Is the WLAN required to support applications such as voice or video streaming, either now or in the future?
Performance expectations	What are the user's performance expectations? If large data files are commonly used, what are the required transfer times?
Areal coverage	What is the operating area in which users will need wireless network coverage? Do usage requirements vary at different locations within this area? Is future growth of the required coverage area expected?
Mobility	If users will move within the operating area while working, will they need to access the WLAN from several fixed locations (roaming) or will they need continuous service while in motion (mobility) — for example to support voice services?
Device interoperability	What types of user devices will need to connect to the network?
User population	What is the total number of users and user devices that are required to be supported? How many users will typically require concurrent service? How much future growth is the network expected to cater for?

Continued

Table 7-1: WLAN User Requirement Types — cont'd

Requirement type	*Considerations*
Security	How confidential is the information transferred across the network? What level of protection is required against unauthorised access?
Battery life	If mobile devices will be used in the network, how often will the user need to recharge battery operated devices?
Economic	What budget is available to implement the WLAN? Are there specific requirements that deliver high value and may justify a higher cost solution?

Table 7-2: WLAN Technical Attributes

Requirement type	*Considerations*
Effective data rate	The required data rate for a single user will be dictated by the usage model, for example by the typical file size and upload/download time, or by the requirements for voice or video streaming. As discussed in Chapter 6, effective data rates can be significantly lower than a standard PHY layer data rate, and will be further affected by adverse environmental factors such as RF interference.
Network capacity	What is the overall network capacity needed to provide the required level of service, given the current and future expected size of the user group and number of user devices? Required capacity will be a key factor both in the technology selection and in determining the appropriate physical architecture for the WLAN.
Quality of service	If the usage model includes applications such as VoWLAN, then guaranteed quality of service will be an important attribute to ensure performance expectations are met.
Application support	Are there specific technical attributes required to support particular usage models?
Network topology	What types of connections are required to meet user requirements? For example, peer-to-peer for local data sharing, point-to-point for linking buildings, etc.
Security	If users' confidentiality requirements are high, then data encryption, network access monitoring and other security measures will be required.

Table 7-2: WLAN Technical Attributes — cont'd

Requirement type	*Considerations*
Interference and coexistence	If the WLAN will have to operate in an environment with other wireless networks, such as Bluetooth, or alongside cordless phones, then coexistence will need to be a consideration.
Technology maturity	Before standards have been agreed early products have an interoperability risk, while a fully mature technology may have limited scope for future development and risk early obsolescence as new usage models arise. The significance of this attribute will depend on whether the user requirements are within the proven capabilities of existing technology or require a leading edge solution.
Operating range	The required range will be determined by the physical extent and nature of the operating area, as well as the layout of components such as access points. The overall link budget will be important in implementing point-to-point connections (wireless bridges between buildings).
Network scalability	If the WLAN is likely to require more than a few access points, or significant future growth is anticipated, then ease of initial configuration and ongoing network management tasks will be a requirement, at least for the network manager.

Establishing Technical Requirements

Technical requirements follow from user requirements, by translating these into the specific technical attributes that are needed to deliver the user requirements (Table 7-2). For example, if there is a user requirement for rapid transfer of large files, for example for video editing applications, this will translate into a technical requirement for a high effective data rate.

Some technical attributes, such as operating range and those relating to interference and coexistence, will be clarified following site surveying and initial planning of the physical layout of the network hardware.

Evaluating Available Technologies

Having established the technical attributes necessary to meet user requirements, the available technologies can then be directly assessed against these attributes. A simple table, similar to the example shown in Table 7-3, can be used to display the assessment, resulting in a transparent and objective comparison of the available solutions.

More sophisticated evaluation methods can also be applied, for example, by assigning a weighting factor to each requirement and a score to each technical solution depending on the extent to which it meets the requirements.

Network Capacity

The total required network capacity will be determined by the sum of the bandwidth requirements of the maximum number of concurrent users expected on the network, with some allowance being made for the fact that this maximum will occur infrequently and some limited degradation of performance may be acceptable during brief periods of high usage. If this requirement exceeds the capacity of a single access point then multiple access points will be required, up to the limit imposed by the number of available non-overlapping channels. The total achievable network capacity for 802.11 networks defined by that limit is shown in Table 7-4.

As described in the Section "Spectrum Management at 5 GHz (802.11h), p. 160", the 802.11h enhancements open up an additional 12 OFDM channels in the 5 GHz band, doubling the achievable network capacity for 802.11a networks.

Operating Range

The operating range of a wireless network link is influenced by a wide range of factors, from the modulation and coding scheme being used to the nature of the materials used in the construction of the building in which the network operates. The key factors are summarised in Table 7-5.

The operating range for 802.11a/b/g networks for varying PHY layer data rates in a typical office environment is shown in Table 7-6, based on transmitted power levels of 100 mW for 802.11a/b, 30 mW for 802.11g and antenna gains of 2 dBi.

Table 7-3: WLAN Technologies; Technical Attribute Comparison

Requirement type	802.11b	802.11g	802.11a	802.11n
PHY layer data rate	11 Mbps	54 Mbps	54 Mbps	200+ Mbps
Effective data rate at MAC SAP	6 Mbps	22 Mbps (8–13 Mbps with 11b stations)	25 Mbps	100 Mbps
Network capacity	3 non-overlapping channels	3 non-overlapping channels	12–24 non-overlapping channels	6–12 non-overlapping dual channels
Quality of service	Not supported	Not supported	Not supported	Supported
Interference and coexistence	2.4 GHz band	2.4 GHz band Interoperable with 802.11b network	5 GHz band	2.4 or 5 GHz band
Technology maturity	Mature	Mature	Mature	Immature
Operating range	Good indoor range including wall penetration	Good indoor range including wall penetration	Line of sight operation. Poor penetration	As for 11b or 11a depending on frequency band
Scalability	Small number of users per AP	Small number of users per AP	Enterprise scale; many users per AP	Enterprise scale; many users per AP

Table 7-4: Effective Network Capacity Comparison for 802.11a, b and g Networks

Network standard and operating mode	MAC SAP rate (Mbps per channel)	Non-overlapping channels	Network capacity (Mbps)
802.11b	6	3	18
802.11g (Mixed mode — RTS/CTS)	8	3	24
802.11g (Non-mixed mode)	22	3	66
802.11a	25	12	300
802.11a (With 802.11h enhancements)	25	24	600

Table 7-5: Factors Affecting WLAN Operating Range

Factor	*Impact on operating range*
Frequency band	As described in the Section "RF Signal Propagation and Losses, p. 112", free space loss is proportional to the logarithm of operating frequency, and increases by 6.7 dB with the increase in frequency from 2.4 to 5.8 GHz.
Transmitter power and receiver sensitivity	These two factors are grouped together since they determine the end points of the link budget, as described in the Section "Link Budget, p. 116"
Modulation and coding scheme	Higher data rate modulation and coding schemes are less robust in that they require correspondingly higher received signal strength to assure accurate decoding. Other things being in equal range therefore decreases for higher data rates.
Environmental factors	Construction materials, particularly metal objects, have a major influence on path loss if RF signals have to pass through walls, ceilings, floors or other obstructions. Path loss is also highly frequency dependent and for all practical purposes a line-of-sight is required for communication in the 5 GHz band.

Table 7-6: WLAN Indicative Indoor Range vs. PHY Data Rate

PHY data rate (Mbps)	*802.11a*	*802.11b*	*802.11g*
54	45		90
48	50		95
36	65		100
24	85		140
18	110		180
12	130		210
11		160	160
9	150		250
6	165		300
5.5		220	220
2		270	270
1		410	410

Selecting the Technical Solution

With the demise of competing standards such as HiperLAN/2 and HomeRF, the essence of the technical choice is simply — which 802.11 flavour best fits the bill?

While the requirements analysis may point to a clear winner among the available technical options, if a selection needs to be made between operating in the 2.4 or 5 GHz bands then an RF site survey, described in the next section, should be conducted as an input to that decision.

Similarly, if the requirements dictate that network capacity needs to be stretched beyond the throughput of a single channel, for example with multiple access points fully exploiting non-overlapping channels, then an initial physical layout may need to be made for both 2.4 and 5 GHz options.

Some on-site physical testing may also be valuable prior to making the final decision, for example to confirm the achievable range if a 5 GHz network is envisaged in a constricted indoor environment.

Future hardware developments may soon make the choice between 2.4 and 5 GHz operation irrelevant. As increasing volumes of dual band radios are shipped, supporting both 802.11g/b and 802.11a, and prices fall to parity with single band products, it will be cost-effective to implement a dual band WLAN that makes the best use of the characteristics of both RF bands.

Planning and Designing the Wireless LAN

Surveying the RF environment

Conducting a site survey is an important step in planning and designing all but the simplest WLAN. It is important to determine the impact of environmental factors on radio wave propagation in the operating area of the LAN, and also to test for the presence of RF signals that will interfere with WLAN performance.

The objective of the site survey is to gather enough information to plan the number and location of access points to provide the required coverage, in terms of achieving a minimum required data rate over the operating area.

There are two types of site survey that can be performed — a noise and interference survey and a propagation and signal strength survey.

Noise and Interference Survey
This survey looks specifically for the presence of radio interference coming from other sources, such as nearby networks, military installations, etc., that could degrade WLAN performance. The main aspects of this type of survey are described below in Table 7-7.

Propagation and Signal Strength Survey
A well executed propagation and signal strength survey will help to ensure that network resources are correctly located so that the planned network will not suffer from coverage holes, resulting in areas of poor network performance, and will also ensure that network capacity is properly planned. The main aspects of this type of survey are described below in Table 7-8.

Table 7-7: Noise and Interference Survey

Survey aspect	*Description*
Objective	Assess the noise floor (RF power per unit of bandwidth, dBm/MHz) across the intended bandwidth to be used by the WLAN. Identify the distribution of RF energy within the bandwidth (frequency, continuous or sporadic, peak and average power levels).
Equipment used	Stand-alone RF spectrum analyser or PC equipped with a wireless interface card and spectrum analysis software.
Survey technique	Site walk-through test prior to access point placement. Extended point measurements should be made at selected locations to identify any intermittent interference sources.
Application of survey results	Noise floor measurements will be used in the link budget calculation to assess the effective range of an RF link given the hardware specifications (transmit power, receiver sensitivity, antenna gains). Interference results will indicate any limitations on bandwidth usage, such as channels that should be avoided, and in extreme cases may dictate a choice between 2.4 and 5 GHz operation.

Table 7-8: Propagation and Signal Strength Survey

Survey aspect	*Description*
Objective	Identify the coverage pattern, received signal strength and achievable data rate for given access point and client station locations within the WLAN operating area.
Equipment used	Laptop or handheld PC equipped with wireless interface card and site survey software. Ideally the survey receiver hardware should be identical to the planned client station hardware (same wireless NIC), otherwise allowances will need to be made, for example for differing receiver sensitivity or antenna gain. Combining the survey tool with a GPS navigation module can help in transferring measurements onto site plans.
Survey technique	Site walk-through test following preliminary access point placement, measuring received signal strength and maximum data rate at each location. If both 2.4 and 5 GHz operation are under consideration, the survey will need to be conducted in each band, as propagation patters will differ widely between these bands.
Application of survey results	Results will show how the ideal omnidirectional propagation pattern is affected by the operating environment (office partitions, walls, cabinets, lift shafts, etc.). Achievable data rate versus range will dictate the required placement of access points as input to planning the physical hardware layout.

Figure 7-1 shows a typical display of propagation and signal strength survey results, which can be used to identify areas where RF propagation is adversely affected by local environmental conditions. Combined with a similar graphical display of noise floor measurements this will give an indication of potential low SNR areas, which may need to be addressed by adjusting access point or antenna locations or by changing transmitter power settings or antenna gain.

Designing the Physical Architecture

Having built up a picture of the RF environment and gathered data on propagation and signal strength in the operating area, a provisional physical layout of the WLAN can be created. The objective of this stage is to establish a layout of hardware that will ensure complete RF coverage and deliver the required bandwidth to wireless client stations.

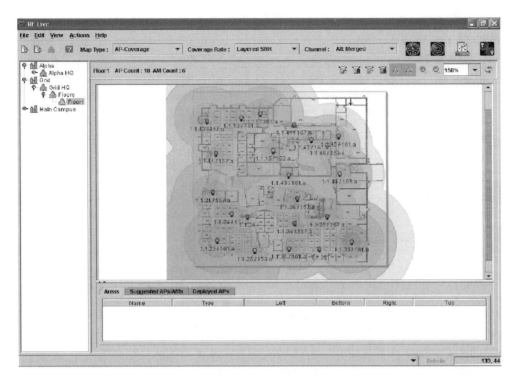

Figure 7-1: Display of Propagation and Signal Strength Survey Results
(Courtesy of Aruba Wireless Networks Inc.)

Network Physical Layout Design

Planning the physical architecture starts with the floor plans of the operating area and the results of the propagation and signal strength survey, and results in a layout plan detailing;

- the required number of access points

- an optimal antenna type and location

- non-conflicting operating channel

- proper power setting for each access point.

Factors influencing physical layout are described in Table 7-9.

The channel allocation patterns shown in Figure 7-2 are based on the three non-overlapping channels 1, 6 and 11 in the 2.4 GHz band. However, in those regulatory domains that permit 13 channels in this band (Europe, Japan, etc., as discussed in the Section "802.11b PHY Layer,

Table 7-9: Factors Influencing WLAN Physical Architecture

Parameter	*Factors influencing the physical architecture*
Number	A preliminary count can be established by dividing the overall operating area per access point by the coverage area determined from the propagation and signal strength survey. The area should be taken out to the contour at which the required data rate could be maintained. The effective coverage area of an access point will be reduced if the propagation pattern is far from omnidirectional as a result of nearby obstacles.
Optimal antenna location	The optimal location for an access point with an omnidirectional antenna will generally be close to the centre of the area to be covered, in a position that maximises line-of-sight to client stations and is clear of obstructions, particularly metal objects such as filing cabinets. An elevated location can be very effective, for example, a ceiling mounted unit.
Operating channel	Any channel that shows significant background or sporadic noise should be avoided. The available non-overlapping channels can then be allocated to access points based on their initial locations. A typical channel allocation pattern for the three non-overlapping channels in the 2.4 GHz band is shown in Figure 7-2.
Power setting	In general, the number of access points will be minimised if maximum permitted transmit power is used. Reasons to adopt a lower power setting may be to reduce out of building propagation or to avoid interference with other RF systems. Conversely high power settings may be required to combat local conditions of high RF noise or high path loss.

p. 152"), it is possible to operate four access points with minimal frequency overlap on channels 1, 5, 9 and 13, potentially increasing network capacity by one-third. This would permit the additional channel allocation patterns shown in Figure 7-3.

Early WLAN layouts were often developed on a trial-and-error basis, by temporarily deploying access points, taking signal strength and throughput measurement (essentially repeating a site survey) and then relocating or adding access points to fill in any identified holes in RF coverage.

Planning tools such as Wireless Valley's "LAN Planner" are available that improve WLAN design by simulation and graphical display of the

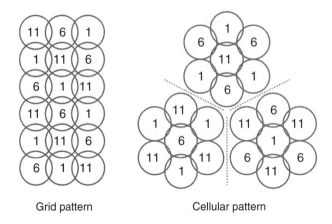

Grid pattern   Cellular pattern

Figure 7-2: Channel Allocation Pattern for 802.11b Access Points with Three Non-overlapping Channels

expected performance of the network. Information such as received signal strength indicator (RSSI), signal to interference ratio (SIR), signal to noise ratio (*SNR*), throughput and bit error rate (*BER*) can be displayed on digitised site plans, allowing more accurate planning in complex environments.

Besides automated placement and configuration of network components, these tools can also ease the later stages of implementation by automatically generating bills of materials and maintenance records.

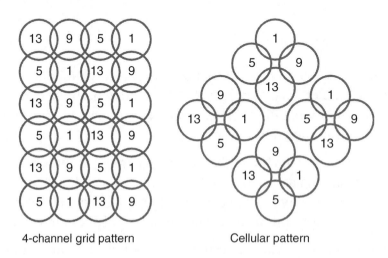

4-channel grid pattern   Cellular pattern

Figure 7-3: Channel Allocation Patterns for Four Non-overlapping Channels

Design and Deployment Using Wireless Switches

Wireless switches, described in the Section "Wireless LAN Switches or Controllers, p. 48", not only change the way in which network traffic is routed between stations, but also provide in-built tools that aid the design and later the configuration process, particularly in large-scale WLAN installations. Typically a wireless switch tool kit will include;

- Automatic layout planning for capacity and coverage (Table 7-10)

- One-click configuration of multiple access points and WLAN switches

- Simple monitoring and operation, from the detection and location of rogue access points to automatic transmitter power adjustment to eliminate coverage gaps and optimise network performance.

Network Bridging — Point-to-point Wireless Links

If the WLAN requirements include linking two separate operating areas, for example providing a link from a WLAN in one building to a wired or wireless LAN in a second, then a point-to-point link will need to be designed.

Since RF propagation is more predictable outdoors than indoors, this will generally be more straightforward than designing indoor layouts as described above. Provided suitable antenna locations can be identified that offer a direct line-of-sight between the locations, a link budget calculation can be performed as described in the Section "Link Budget, p. 116".

Table 7-10: Wireless Switch Automated Planning Tools

Wireless switch feature	Description
Automated site survey tools	Most wireless switches provide automated site survey tools that generate input survey data for simulation of the WLAN operating environment.
Access point location planning	Tools allow import of building blueprints and construction specifications from CAD programs and in-built propagation models are used to determine optimal access point location, taking account of construction materials, path losses and attenuation.

The required combination of transmit power and antenna gain needed to deliver sufficient received signal strength to achieve the desired data rate can be determined, given the link range and intended RF band.

Pilot Testing

The design process described above will have established a layout of WLAN hardware which aims to deliver the required data throughput with complete RF coverage of the operating area. A pilot test of the design solution, prior to complete installation, will be helpful to ensure that no requirements have been missed and no limitations of the selected technology have been overlooked. Aspects of pilot testing are summarised in Table 7-11.

A pilot test will involve the installation of a number of access points of the same type as are intended for the final installation, to cover part of the operating area. Selecting the part of the operating area that posed the greatest difficulty in the design stage — either because of interference or identified propagation issues — will provide the most robust challenge to the design solution.

The pilot test should include monitoring and user responses to the day-to-day performance of the pilot installation, as well as stress testing — running under extreme load conditions.

The results of the pilot test will either validate the installation proposed at the design stage, or they may indicate that adjustments in the design are necessary if user requirements and performance expectations are to be fully met. A repeat pilot testing may be valuable if major changes in the design solution are indicated.

Installation and Configuration

WLAN Installation
Installation should follow a systematic approach in which each building block of the WLAN is installed and tested before moving on to the next. This will ensure that any problems are identified as early as possible — once the whole network is in place it may be much more difficult to identify a single faulty element.

Table 7-11: Aspects of WLAN Pilot Testing

Pilot test feature	*Description*
Stress testing	Stress testing of the pilot installation will involve loading the wireless LAN with the most demanding data transfer requirements, such as video or voice streaming or transfers of large files. Gradually increasing the number of concurrent users of high rate applications will test the achievable throughput limit.
Maintain and analyse user logs	Access point logs should be enabled during the pilot test period, and analysed to show peak usage times and identify the types of services that were used during the pilot. A comparison with the stated user requirements will test whether the expected usage patterns are realistic.
Conduct post-pilot user surveys	User surveys conducted after the pilot has been in operation for some time can highlight the type and frequency of any problems, and test whether user expectations are being met. Problem reports can be matched to access point logs to highlight any bottlenecks in the pilot installation that may point to necessary design changes before full-scale implementation.

The key implementation steps are described in Table 7-12. If the installation involves multiple access points, follow these steps for each BSS in turn — verifying operation of each before moving on to the next.

After installing the first access point, Steps 4–6 can be followed for each station to be connected to the access point. Steps 1–6 can then be followed for the second and any additional access points.

WLAN Configuration

The initial configuration will ensure that WLAN parameters are set-up to enable communication between the installed access points and stations, and ensure the coexistence of adjacent BSSs. After completing the configuration of access points and stations, the network operating system may also need to be configured.

Table 7-12: WLAN Implementation Steps

Implementation step	Description/Considerations
(1) Install Ethernet cabling to the planned access point location	If access point locations have not been pilot tested it may be desirable to lay temporary cabling until the access point locations have been verified in practice, especially if cable laying is expensive and disruptive. Test the cabling by connecting a laptop to the wired network via the new cable and "ping" testing connectivity (see the Section "Analysing Wireless LAN Problems, p. 241").
(2) Install the first access point at its planned location and connect to the wired network	Follow the vendor's instructions and use any mounting kit provided, for example, for wall, ceiling or above suspended ceilings. Connect network cables and antenna(s), following vendor recommendations regarding antenna orientation. Connect power cable unless power-over-Ethernet is being used.
(3) Configure the access point settings	Configuration is described further in the following section. Depending on the hardware in use, it may be necessary to install configuration software on a computer connected to the wired network, or to configure the access point via a web based utility.
(4) Install wireless NICs in the computers that will connect via this access point	Check vendors instruction to determine the correct installation sequence (drivers first or hardware first), which is often operating-system dependent (e.g. hardware first for Windows XP).
(5) Configure a wireless network connection for each new station	After installation of software drivers, the wireless NIC may need to be identified to the computer's operating system as a new network connection.
(6) Verify operation of all stations	Confirm network connectivity for stations connected through the first access point.

Access Point Configuration

The details of access point configuration will depend on the specific hardware selected, but Table 7-13 gives a summary of the basic configuration parameters that will be applicable in all cases.

If the access point is a dual- or tri-mode device, with multiple radios, each will require separate configuration. Some access points may also allow other PHY layer parameters to be configured, such as the fragmentation

Table 7-13: Access Point Configuration Parameters

Parameter	*Configuration considerations*
IP address (plus subnet mask and default gateway)	The access point may have a default IP address, defined by the manufacturer, that will enable a direct connection through a web browser. Alternatively an IP address may be assigned to the access point by a DHCP server in the wired network or a static IP address (plus subnet mask and default gateway) may be assigned by connecting a PC to the access point's configuration port.
SSID	The service set ID should be changed from its default value as a basic security measure in every network. A policy for SSID assignment may be defined for large WLAN installations.
SSID broadcast	SSID broadcast within beacon frames can be either enabled or disabled. Disabling SSID broadcast is a further security measure, discussed in Chapter 8.
Maximum transmit power	Setting the maximum allowable transmit power level to comply with local regulations.
Radio channel	Selection of the operating channel within the range allowed by local regulations. Some access points may include an automatic search for the least congested channel.
Operating mode	In the case of 802.11g networks, a selection can be made between mixed mode and g-only mode, depending on whether 802.11b stations will also be operating in the network.
Security	Selection of security modes (64-bit WEP, 128-bit WEP, WPA-PSK, etc.) and entry of passphrase or encryption keys, transmit key and authentication mode (described further in Chapter 8).
Antenna configuration	An access point with multiple antennas may be configured to use one specified antenna, or to use all antennas in diversity mode — selecting the antenna that gives the strongest signal.

Figure 7-4: Typical Access Point Configuration Screens

limit or the maximum number of retries before a packet is dropped. Figure 7-4 shows typical set-up screens for configuring access point operating parameters.

Wireless Station Configuration

As for access points, station configuration details will be hardware dependent, but Table 7-14 describes the basis wireless NIC configuration

Table 7-14: Wireless NIC Configuration

Parameter	Configuration considerations
Network mode	Select infrastructure mode if the station will be associating with an access point, or ad-hoc mode if the station will only be connecting to other stations and not to an access point.
SSID	Enter the SSID of the access point that this station will be connecting to.
Radio channel	Selection of the operating channel within the range allowed by local regulations.
Operating mode	As for access points, for 802.11g hardware a selection can be made between mixed mode and g-only mode in 802.11g networks.
Security	Security settings (passphrase or encryption keys, transmit key and authentication mode) must match those entered for the access point that the station will be associating with.

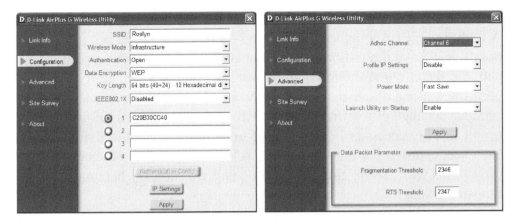

Figure 7-5: Typical Wireless NIC Configuration Screens

parameters that will be applicable in all cases. Figure 7-5 shows typical set-up screens for configuring wireless NIC operating parameters.

Network Operating System Configuration
If the WLAN is an extension of an existing wired network, then no additional network operating system (NOS) configuration will be required. However, if the WLAN is creating a new network, a number of NOS configuration tasks will also have to be completed. Details will depend on the specific NOS in use, but typical tasks will include;

- ensuring that network protocols such as TCP/IP are installed

- ensuring that networking software is installed e.g. for file and printer sharing

- identifying workgroups of users that share resources

- enabling parts of the network file system for common or workgroup access

- enabling devices such as printers, scanners, etc., for shared access.

Automatic WLAN configuration and management

A traditional WLAN deployment, based on first generation or "fat" access points, will have limited or no built-in network management capabilities, and initial configuration and ongoing management will generally be performed using a web-based user interface. Network management tasks, such as a change of security settings, RF operating channel, transmit power level or access policy, will have to be implemented individually on each access point. For a corporate WLAN, as the number of access points grows, this will be very time consuming and will quickly become unmanageable.

Second generation WLAN hardware, in the form of wireless switches, are designed to enable these management tasks in WLANs with many access points, and provide a range of automated configuration and management tools, as summarised in Table 7-15.

Table 7-15: Wireless Switch Automated Configuration and Management Tools

Wireless switch feature	*Description*
Automatic configuration	Once access points are deployed, wireless switches can provide automatic configuration by determining the best RF channel and transmit power settings for individual access points, reducing implementation time and the risk of errors inherent in manual configuration.
Access control	Access points can be grouped according to categories such as the building or floor where they are located, and lists can be constructed that specify which access points or groups specific clients are permitted to connect to. Access control can include monitoring a station's roaming history and bandwidth usage.
RF management	Some wireless switch products can continuously adapt to changes in the RF environment by changing RF operating channels and power settings in order to avoid channels affected by noise and interference.
Enhanced security	RF site surveys can be run to detect and locate rogue access points and unauthorised users or ad-hoc networks.

Operation and Support

With the completion of installation and configuration, attention will turn to the operation and support of the WLAN. Initial user training, day-to-day help desk support and hardware maintenance will need to be planned and provided, depending on the size of the user group. Two key tasks to assure that the installation continues to meet user requirements are network performance monitoring and the control of future changes in the network installation or configuration.

Network Performance Monitoring

For medium to large-scale WLAN implementations, the network manager will need to keep an overview of network performance in order to identify and diagnose the nature and location of any problems, and quickly determine a solution.

A variety of WLAN management tools are available to ease this performance monitoring task. Typically a graphical interface, based on building blueprints or other plans of the operating area, allows the network manager to view performance data collected from access points and interface cards.

This real time performance data, collected using SNMP queries to access points and stations, as well as from system software, will be able to identify;

■ Active access points and client stations

■ Average data rates, retry rates and overall network utilisation

■ Noise levels and interference by area

■ Network areas or individual access points experiencing significant error rates.

Aruba Network's RF Live™ is an example of this type of WLAN management software. A view of the real time management dashboard is shown in Figure 7-6.

Network Change Control

Having set-up the WLAN based on a well thought out physical layout and configuration plan, future changes must be given the same degree of

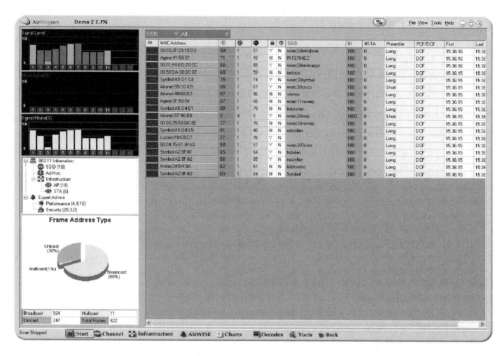

Figure 7-6: WLAN Management Software Display (Courtesy of AirMagnet Inc. © 2006 AirMagnet ® is a registered trademark.)

thought and planning if the effectiveness of the network is to be maintained. The addition of an access point, perhaps aimed at filling in a coverage gap or providing additional network capacity for a new concentration of users, could easily have the opposite effect on performance if the new hardware is not configured for coexistence with the existing set-up, with appropriate channel selection, transmit power and security settings.

A documented set of network policies and a supporting change control procedure will ensure that future change is managed in such a way that performance is maintained or enhanced. Network policies may address issues such as;

- Supported standards (e.g. 802.11b/g)

- Supported hardware vendors or NIC models

- SSID management

- Security and encryption requirements (e.g. personal firewalls, WEP/WPA).

A change control procedure should define how proposed changes to the installation, including hardware, software and configuration settings, will be technically reviewed and endorsed prior to implementation. The procedure should cover;

- Scope of the procedure — what types of changes are controlled

- Process and documentation required for proposing changes

- Roles and responsibilities of reviewers and decision makers

- Authorities required to endorse different types of changes.

A Case Study: Voice over WLAN

In order to meet the demands of supporting voice applications, a wireless LAN needs to fulfil stringent performance requirements, particularly if the provision of voice services was not a consideration during the initial design. Close attention will need to be paid to wireless coverage, quality of service and seamless roaming, recognising that the roaming area for voice coverage is likely to be different from the operating area for other roaming coverage such as laptop connections.

VoWLAN Bandwidth Requirements

Voice usage puts high demands on WLAN bandwidth, and an estimate of the number of concurrent calls that need to be supported will be a key factor in determining either the required design or the suitability of an existing WLAN design.

Although a single voice session requires around 64 kbps of bandwidth, or as little as 10 kbps with compression, IP and 802.11 MAC protocol overheads increase the required bandwidth to around 200 kbps per session. Collisions on the shared wireless medium will further limit the number of concurrent voice sessions. The actual number of concurrent calls that can be supported by a single access point, while providing voice quality comparable to a toll call, will depend on the wireless standard in use, as shown in Table 7-16.

The concurrent call limits noted here are indicative only, and apply to a network carrying voice traffic alone. If the network is carrying data

Table 7-16: VoWLAN Capacity of 802.11 WLANs

Standard	PHY data rate (Mbps)	MAC SAP data rate (Mbps)	Maximum concurrent voice calls
802.11b	11	6	6–7
802.11b + g	54	9–15	7–8
802.11g	54	22	ca. 20
802.11a	54	25	ca. 25

as well as voice traffic, call quality will be affected and the maximum number of concurrent calls will drop.

As these capacity limits are approached, load balancing will be required to ensure that access points do not become overloaded. Access points alone will not have the overview of the available infrastructure or total traffic to perform this function, but a wireless switch will be able to monitor the total number of voice sessions being routed at any moment and intelligently manage capacity by handing off VoWLAN traffic where possible to alternative access points.

RF Coverage

Voice services put greater demands on wireless coverage, so that roaming users do not encounter coverage gaps in corridors or stair wells while talking on their VoWLAN phones. If the WLAN has been designed to provide coverage of work areas and meeting rooms, a more focussed propagation and signal strength survey will be required when planning for VoWLAN service.

Quality of service to voice clients will be improved if only the maximum supported data rate is enabled on all access points. This ensures that overall network throughput is not slowed when a client moves away from the access point, but also means that a higher density of access points will be necessary to give adequate RF coverage to support voice traffic.

Quality of Service

VoWLAN requires guaranteed quality of service in order to minimise packet loss, delay and jitter that cause degradation of voice quality. As described in the Section "Quality of Service (802.11e specification), p. 157", the 802.11e standard has been developed to address the lack of QoS in the original 802.11 specification, and the Wi-Fi Alliance has also released wireless multimedia (WMM) as an interim subset of the 802.11e specification.

WMM and 802.11e prioritise voice traffic over video, best effort and background traffic categories, and 802.11e or WMM compliance will be an essential technical requirement for WLAN hardware in all but the least demanding VoWLAN applications.

However, 802.11e and WMM define traffic categories by device and not by application, so that frames transmitted by a laptop user running a software phone will be queued according to the device priority, which will typically be best effort. Future enhancements of QoS standards are likely to provide for prioritisation by application.

Seamless Roaming

Complete RF coverage alone is not sufficient to ensure the seamless roaming required for uninterrupted VoWLAN service. Roaming VoWLAN clients also need to make fast transitions between access points, avoiding latency and packet loss during hand-off, so that service quality is not degraded or interrupted.

When a voice client transitions from one access point to another, association and authentication at the new access point must be fast to avoid degradation of voice quality. Typically, a VoWLAN performs optimally with packet delays and jitter of under 50 ms, and a call will drop out when delays approach 150 ms. By comparison, associating and authentication with an access point typically takes from 150 to 500 ms, and measured roaming times for VoWLAN clients can be anything from 1 to 4 seconds with multiple clients making concurrent transitions. To achieve the faster transitions required for seamless roaming, pre-authentication must be completed prior to hand-off, as provided for in the 802.11r standard.

Dual-mode handsets are now available that allow roaming between cellular phone networks and WLANS, with the access point or wireless switch controlling the hand-off between the VoWLAN service and a cellular phone service.

Pilot Testing

As described in Chapter 6, a number of upcoming 802.11 enhancements are aimed at improving RF management, QoS and roaming in order to improve the capability of 802.11 networks to support VoWLAN services. Until these standards are ratified and published, and compliant products become available, VoWLAN deployments will require careful pilot testing to ensure that the required service can be delivered.

Stress testing should be conducted by loading the WLAN with multiple concurrent calls and testing voice quality with and without background data streams being present on the network. Coverage and roaming ability can also be assessed using modelling tools, but ultimately the quality of voice services can be best confirmed by pilot testing.

VoWLAN Security

When using a conventional telephone system, physical access to telephone lines or to a private branch exchange (PBX) in an office is required in order to intercept voice traffic. The physical security of these systems means that the encryption of voice traffic over a conventional telephone system is only justified for highly security-sensitive organisations.

In the case of VoWLAN, voice traffic going outside the organisation will pass over the completely unsecured Internet, and it becomes relevant for many more organisations to provide the same degree of security for voice as for data traffic. VoWLAN services will be susceptible to the range of security attacks described in Chapter 8, and will be particularly sensitive to Denial of Service attacks in view of the delay sensitive nature of voice traffic. Specific VoWLAN/VoIP security considerations are discussed in the Section "VoWLAN and VoIP Security, p. 239".

A further security concern for VoWLAN/VoIP services is the possibility of unsolicited bulk messages being broadcast to Internet connected phones. As IP telephony becomes more widespread, so called spam over internet telephony (SPIT) could become as pervasive as spam e-mail. Technology to counter this threat is under development, based either on filtering and deleting unwanted calls based on characteristics such as call frequency and duration, or identifying authentic calls by enabling these to be digitally signed and checking signatures against an authorised caller list.

Wireless LAN Security

The Hacking Threat

The flexibility of wireless networking is bought at the price of an increased need to think about security. Unlike a wired network, where signals are effectively limited to the connecting cables, WLAN transmissions can propagate well beyond the intended operating area of the network, into adjacent public spaces or nearby buildings. With an appropriate receiver, data transmitted over the WLAN may be accessible to anyone within range of the transmitter.

Although the spread spectrum modulation technologies described in Chapter 4 were originally conceived for military applications where jamming was the main concern, any device conforming to a wireless standard such as 802.11b will be able to intercept wireless data traffic from a WLAN operating that standard. Additional security measures must be taken to prevent unauthorised access to user data and network resources.

The easy accessibility of wireless networks resulted in a new industry springing up among the early adopters to map out networks and make others aware of the opportunity for free, although often illegal, access. War Driving and War Walking describe the pursuits of driving or walking around with a wireless enabled laptop or handheld device and seeking out wireless networks.

War Chalking sprang up in London in 2002 as a way of making others aware of nearby wireless networks that could be freely accessed. Some WLANs are deliberately left without security measures enabled in order to allow

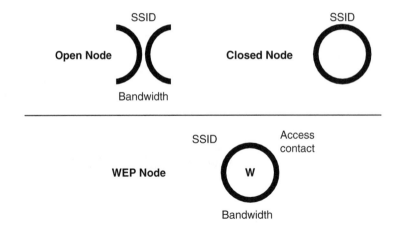

Figure 8-1: War Chalking; Table of Symbol Elements and Captions

free public access, and War Chalking symbols were intended to help people living in or visiting a city to identify these networks in order to connect to the Internet (Figure 8-1). The site www.warchalking.org gives more information, as well as an interesting discussion on the legal and moral aspects of accessing unsecured WLANs.

These free access points are of no interest to the determined hacker (or cracker), who is motivated by the challenge of breaking into a secured network. The measures described later in this chapter should be followed to ensure that access to a wireless network is as secure as it can be, and to minimise this risk of unauthorised access and use.

At home, it is not just deliberate hackers who might see an unsecured wireless network as a free resource. If basic security measures are not enabled, any wireless enabled computer in a neighbouring house or apartment will be able to connect to the network and make free use of resources such as an Internet connection.

Wireless LAN Security Threats
The type of security threats faced by a wireless LAN are many and varied, and although initially targeted at the PHY and MAC layers, the ultimate goal is to access or disrupt data or activity at the application layer. A few of the main vulnerabilities are described below.

- Denial of service (DoS) attacks – an attacker floods a network device with excessive traffic, preventing or seriously slowing normal access. This can be targeted at several levels, for example, flooding a web server with page requests or an access point with association or authentication requests.

■ Jamming – a form of DoS in which an attacker floods the RF band with interference, causing WLAN communication to grind to a halt. In the 2.4 GHz band this could be done using Bluetooth devices, some cordless phones or a microwave oven!

■ Insertion attacks – an attacker is able to connect an unauthorised client station to an access point, either because no authorisation check was made or because the attacker masqueraded as an authorised user.

■ Replay attack – an attacker intercepts network traffic, such as a password, and uses it at a later time to gain unauthorised access to the network.

■ Broadcast monitoring – in a poorly configured network if an access point is connected to a hub rather than a switch, the hub will broadcast data packets that may not be intended for wireless stations, and these can be intercepted by an attacker.

■ ARP Spoofing (or ARP cache poisoning) – an attacker can trick the network into routing sensitive data to the attacker's wireless station, by accessing and corrupting the ARP cache in which MAC and IP address pairs are stored.

■ Session hijacking (or man-in-the-middle attack) – a type of ARP spoofing attack in which an attacker breaks a station's connection with the access point, by posing as the station and disassociating itself, and then poses as the access point to get the station to associate with the attacker.

■ Rogue access point (or evil twin intercept) – an attacker installs an unauthorised access point with the correct SSID (the twin). If the signal is strengthened using an amplifier or high gain antenna, clients stations will preferentially associate with the rogue access point and sensitive data will be compromised.

■ Cryptoanalytic attacks – an attack in which the attacker uses a theoretical weakness to break the cryptographic system. An example is the weakness of the RC4 cipher that leads to the vulnerability in WEP (see the Section "WEP Wired Equivalent, Privacy Encryption, p. 209").

■ Side channel attacks – an attack in which the attacker uses physical information, such as power consumption, timing information or acoustic or electromagnetic emissions to gain information about the cryptographic system. Analysis of this information might allow the attacker to determine an encryption key directly or a plaintext message from which the key can be computed.

Although the range of threats is wide and varied, in most cases executing these kinds of attack requires a high level of technical expertise on the part of the hacker. The risk to network security can be significantly reduced by enabling the full range of available security measures described in the following sections.

WLAN Security

Table 8-1 summarises the range of generic security measures that have been developed to protect wireless LANs from the threats and vulnerabilities described above.

As described in the Section "WEP – Wired Equivalent Privacy Encryption, p. 209", the original 802.11 standard included only limited

Table 8-1: Wireless LAN Security Measures

Security measure	*Description*
User authentication	Confirms that users who attempt to gain access to the network are who they say they are.
User access control	Allows access to the network only to those authenticated users who are permitted access.
Data privacy	Ensures that data transmitted over the network is protected by encryption from eavesdropping or any other unauthorised access.
Key management	Creation, protection and distribution of keys used for encrypting data and other messages.
Message integrity	Checks that a message has not been modified during transmission.

authentication and weak encryption. The interim development and deployment of enhancements to 802.11 security has been led by the Wi-Fi Alliance with the release of WPA and WPA2.

The shortcomings of the original 802.11 standard were addressed with the ratification in 2004 of the 802.11i standard, which provides a standards basis for WPA and WPA2. The progressive enhancements to WLAN security, and the technologies underlying these enhancements, are described in the Sections "WEP – Wired Equivalent Privacy Encryption, p. 209", "Wi-Fi Protected Access; WPA, p. 212" and "IEEE 802.11i and WPA2, p. 219", and the Sections "WLAN Security Measures, p. 230" and "Wireless Hotspot Security, p. 236" then return to the practical aspects of ensuring security in wireless LANs, at wireless hotspots and some specific aspects of VoWLAN security.

WEP – Wired Equivalent Privacy Encryption

As the name suggests, the intention of WEP was to provide a level of security equivalent to a wired network, although this aspiration was not achieved because of a fundamental cryptographic weakness. From the list of security measures summarised in Table 8-1, WEP provides a limited degree of access control and data privacy using a secret key, typically a passphrase, that is entered into the access point and is required to be known by any station attempting to associate with the access point. Without knowledge of the passphrase, a station will be able to see network traffic but will not be able to associate or easily decrypt data.

WEP encryption translates the passphrase into a 40-bit secret key, to which a 24-bit initialisation vector (IV; see Glossary) is added to create a 64-bit encryption key. As an interim attempt to strengthen WEP encryption, some vendors enhanced the key length to 128-bit (104-bit + 24-bit IV). This proved to be largely a cosmetic enhancement since, as described below, the underlying vulnerability meant that an eavesdropper could still derive the key by analysing roughly 4 million transmitted frames, whether 40-bit or 104-bit keys were used.

The input data stream, known in cryptographic terminology as the plaintext, is combined with a pseudo-random key bit stream in an XOR (exclusive OR) operation to create the encrypted ciphertext. WEP creates the key bit stream using the RC4 algorithm to make a pseudo-random

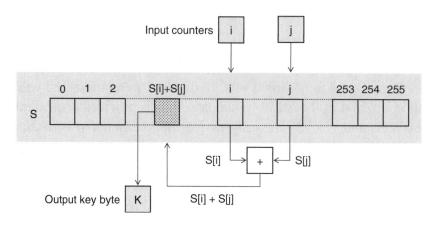

Figure 8-2: Key Stream Generation Using the RC4 Algorithm

selection of bytes from a sequence S that is a permutation of all 256 possible bytes.

As shown in Figure 8-2, RC4 selects the next byte in the key stream by;
    Step (1) incrementing the value of a counter i,
    Step (2) incrementing a second counter, j, by adding the value of $S(i)$, the i'th byte in the sequence, to the previous value of j,
    Step (3) looking up the values of the two bytes $S(i)$ and $S(j)$, indexed by the two counters, and adding them together modulo 256,
    Step (4) outputting the byte K indexed by $S(i) + S(j)$,
        i.e. $K = S(S(i) + S(j))$.

The values of bytes $S(i)$ and $S(j)$ are then exchanged before returning to step (1) to select the next byte in the key stream.

The initial permutation of bytes in S is determined by a key scheduling algorithm which uses a similar manipulation of bytes within S, starting with the identity permutation of bytes (0, 1, 2, 3, 4, ..., 255), but at each Step (2), when the counter j is incremented, a byte from the 64-bit or 128-bit encryption key is also added to the counter.

WEP also provides limited message integrity checking using a cyclic redundancy check (CRC-32; see Glossary) to compute a 32-bit integrity check value (ICV) which is appended to the data block before encryption. The full WEP computational sequence is then as follows (see Figure 8-3);

    Step (1) the ICV is computed for the data block to be transmitted in the frame

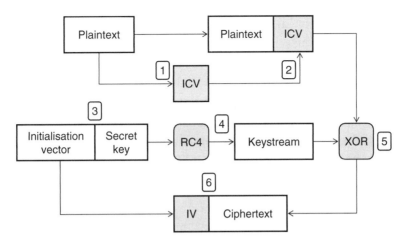

Figure 8-3: WEP Encryption Process

Step (2) the ICV is appended to the data block

Step (3) The initialisation vector is combined with the secret key to generate the full encryption key

Step (4) the RC4 algorithm is used to transform the encryption key into a key stream

Step (5) an XOR operation is performed between the key stream and the output of Step (2)

Step (6) the initialisation vector is combined with the cipher text.

Although WEP provides a reasonable level of security against casual eavesdroppers, its weaknesses were recognised soon after its release as part of the 802.11 standard. In 2001, cryptographers Scott Fluhrer, Itsik Mantin and Adi Shamir realised that, because of a weakness in the RC4 key scheduling algorithm, the output key stream was significantly non-random. This allows the encryption key to be determined by analysing a sufficiently large number of data packets encrypted using the key.

In essence, WEP transmits information about the encryption key as part of the encrypted message so that a determined hacker, equipped with the necessary tools, could collect and analyse transmitted data to extract the encryption key. This requires several million packets to be intercepted and analysed, but could still be accomplished in under an hour on a high traffic network. WEP also uses a static shared key, as there is no mechanism for changing the key other than manual re-entry of a new key or passphrase into every device that operates on the WLAN.

It was not technological limitations that limited the strength of the encryption algorithm used in 802.11 but, interestingly, US export controls. The export of data encrypting technology was considered by the US government to be a threat to national security and as a result the WEP scheme could not be the strongest available if it was to be adopted as an international standard. These restrictions have since been lifted and new, more powerful encryption methods have been developed, such as the Advanced Encryption Standard (Section "IEEE 802.11i and WPA2, p. 219").

Wi-Fi Protected Access — WPA

To overcome these known vulnerabilities in the original 802.11 security implementation, the Wi-Fi Alliance developed Wi-Fi Protected Access (WPA) as a means to provide enhanced protection from targeted attacks. WPA was an interim measure that was based on a subset of the enhanced security mechanisms that were then still under development by 802.11 TGi as part of the 802.11i standard.

WPA uses the temporal key integrity protocol (TKIP) for key management, and offers a choice of either the 802.1x authentication framework together with extensible authentication protocol (EAP) for enterprise WLAN security (Enterprise mode), or simpler pre-shared key (PSK) authentication for the home or small office network which does not have an authentication server (Personal mode).

These measures, which were initially available as firmware upgrades to Wi-Fi compliant devices, first came to market in early 2003. In 2004, a further strengthening of the encryption was introduced in the second generation WPA2. This replaced RC4, still used in WPA, with the advanced encryption standard (AES) which was ratified as part of the 802.11i standard in June 2004. The key components of WPA and WPA2 are described in the following sections.

Temporal Key Integrity Protocol

The WEP encryption vulnerability was addressed in WPA by two new MAC layer features: a key creation and management protocol called TKIP (temporal key integrity protocol) and a message integrity check function (MIC). Features of WEP and WPA key management are compared in Table 8-2.

Table 8-2: WEP and WPA Key Management and Encryption Compared

Security feature	WEP	TKIP
Temporal key/Passphrase	40-bit, 104-bit	128-bit
Initialisation vector (IV)	24-bit	48-bit
Keys	Static	Dynamic
Encryption cipher	RC4	RC4

After a station has been authenticated, a 128-bit temporal key is created for that session, either by an authentication server or derived from a manual input. TKIP is used to distribute the key to the station and access point and to set up key management for the session. TKIP combines the temporal key with each station's MAC address, plus the TKIP sequence counter, and adds a 48-bit initialisation vector to produce the initial keys for data encryption.

With this approach each station will use different keys to encrypt transmitted data. TKIP then manages the update and distribution of these encryption keys across all stations after a configurable key lifetime that might be from once every packet to once every 10,000 packets, depending on security requirements. Although the same RC4 cipher is used to generate an encryption key stream, TKIP's key mixing and distribution method significantly improves WLAN security, replacing the single static key used in WEP with a dynamically changing choice from 280 trillion possible keys.

WPA supplements TKIP with a message integrity checking (MIC) that determines whether an attacker has captured, altered and resent data packets. Integrity is checked by the transmitting and receiving stations computing a mathematical function on each data packet.

While the simple CRC-32, when used to compute the ICV in WEP, is adequate for error detection during transmission, it is not sufficiently strong to assure message integrity and prevent attacks based on packet forgery. This is because it is relatively easy to modify a message and re-compute the ICV to conceal the changes. In contrast, MIC is a strong cryptographic hash function, which is calculated using source and destination MAC addresses, input data stream, the MIC key and the TKIP sequence counter (TSC).

If the MIC value computed by the receiving station does not match the MIC value received in the decrypted data packet, the packet is discarded

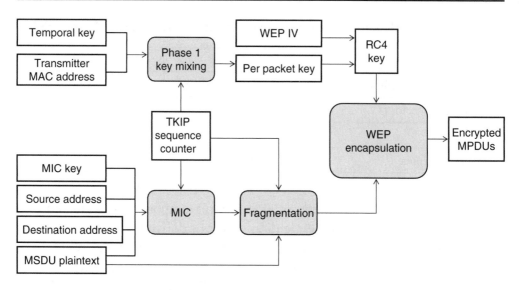

Figure 8-4: TKIP Key Mixing and Encryption Process

and countermeasures are invoked. These countermeasures consist of resetting keys, increasing the rate at which keys are updated, and sending an alert to the network manager. MIC also includes an optional countermeasure, which will deauthenticate all stations and shutdown the BSS for any new association for one minute, if an access point receives a series of altered packets in quick succession. The complete WPA encryption and integrity checking process is shown in Figure 8-4.

802.1x Authentication Framework

IEEE 802.1x is an access control protocol that provides protection for networks by authenticating users. After successful authentication, a virtual port is opened on the access point for network access, while communications are blocked if authentication fails. 802.1x authentication defines three elements;

- The Supplicant – software running on the wireless station that is seeking authentication

- The Authenticator – the wireless access point that requests authentication on behalf of the supplicant and

- The Authentication Server – the server, running an authentication protocol such as RADIUS or Kerberos, that provides centralised authentication and access control using an authentication database.

The standard defines how the extensible authentication protocol (EAP) is used by the Data Link layer to pass authentication information between the supplicant and the authentication server. The actual authentication process is defined and handled depending on the specific EAP type used, and the access point, acting as an authenticator, is simply a go-between, enabling the supplicant and the authentication server to communicate.

Authentication Servers (RADIUS)

The application of 802.1x authentication in an enterprise WLAN, requires the presence of an authentication server within the network, which can authenticate users against a stored list of the names and credentials of authorised users. The most commonly used authentication protocol is the remote authentication dial-in user service (RADIUS), which is supported by WPA compliant access points and provides centralised authentication, authorisation and accounting services.

To authenticate a wireless client seeking network access via an access point, the access point, acting as a client to the RADIUS server, sends a RADIUS message to the server which contains the user's credentials together with information on the requested connection parameters (Figure 8-5). The RADIUS server will either authenticate and authorise or reject the request, in either case sending back a response message.

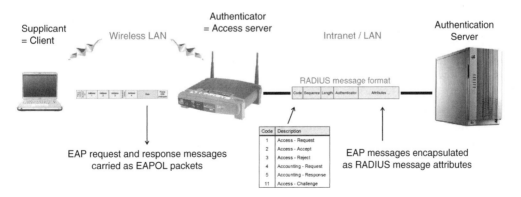

Figure 8-5: Message Format for EAP Over RADIUS Authentication

A RADIUS message comprises a RADIUS header and RADIUS attributes, with each attribute specifying a piece of information about the requested connection. For example, an Access-Request message will contain attributes for the user name and credentials, and the type of service and connection parameters being requested by the user, while the Access-Accept message contains attributes for the type of connection that has been authorised, relevant connection constraints and any vendor specific attributes.

Extensible Authentication Protocol
The extensible authentication protocol (EAP) builds on the framework for enabling remote access that was originally established for dial-up connections in the point-to-point protocol (PPP) suite of protocols.

The PPP dial-up sequence provided for the negotiation of link and network control protocols, as well as the authentication protocol that would be used, based on the desired level of security. For example, an authentication protocol, such as password authentication protocol (PAP) or challenge-handshake authentication protocol (CHAP), is negotiated between client and the remote access server when a connection is established and then the chosen protocol is used to authenticate the connection.

EAP extended this structure by allowing the use of arbitrary authentication mechanisms, called EAP types, which define various structures for the authentication message exchange. When a WLAN connection is being established, client and access point agree on the use of EAP for authentication, and a specific EAP type is chosen at the start of the connection authentication phase. The authentication process then consists of the exchange of a series of messages between the client and authentication server, the length and detail of the exchange depending on the requested connection parameters and the selected EAP type. Some of the most common EAP types are described below.

When EAP is used together with RADIUS as the authentication protocol, EAP messages sent between the access point and the authentication server will be encapsulated as RADIUS messages, as shown in Figure 8-5.

Extensible Authentication Protocol over LANs
To apply EAP to LANs or WLANs rather than to dial-up connections, extensible authentication protocol over LAN (EAPoL) was defined in the

802.1x standard as a transport protocol for delivering authentication messages. EAPoL defines a set of packet types that carry authentication messages, the most common of which are;

- EAPoL-Start – Sent by the authenticator to start an authentication message exchange

- EAP-Packet – Carries each EAP message

- EAPoL-Key – Carries information related to generating keys

- EAPoL-Logoff – Informs the authenticator that the client is logging off.

EAP Types

EAP types supported by the Wi-Fi Alliance's interoperability certification programme include; EAP-TLS, EAP-TTLS/MS-CHAP v2, PEAP v0/EAP-MS-CHAP v2, PEAP v1/EAP-GTC and EAP-SIM. To give a flavour of how these EAP types differ, EAP_TLS, EAP-TTLS and PEAP are briefly described here.

EAP-TLS (Transport layer security) uses certificate based authentication between client and server, and can also dynamically generate keys to encrypt subsequent data transmissions.

An EAP-TLS authentication exchange requires both the station and the authentication (RADIUS) server to prove their identities to each other using public key cryptography and the exchange of digital certificates (see next section). The client station validates the authentication server's certificate and sends an EAP response message that contains its certificate and starts the process of negotiating encryption parameters, such as the cipher type that will be used for encryption. As shown in Figure 8-6, once the authentication server validates the client's certificate, it responds with the encryption keys to be used during the session.

EAP-TLS therefore requires initial configuration of certificates on both the client station and the authentication server, but once this is established by the network manager no further user intervention is required. On the client station, the certificate must be protected by a passphrase or PIN, or stored on a smart card. The result is a very high level of wireless security although, for large WLAN installations, the requirement to manage both client and server certificates as well as the

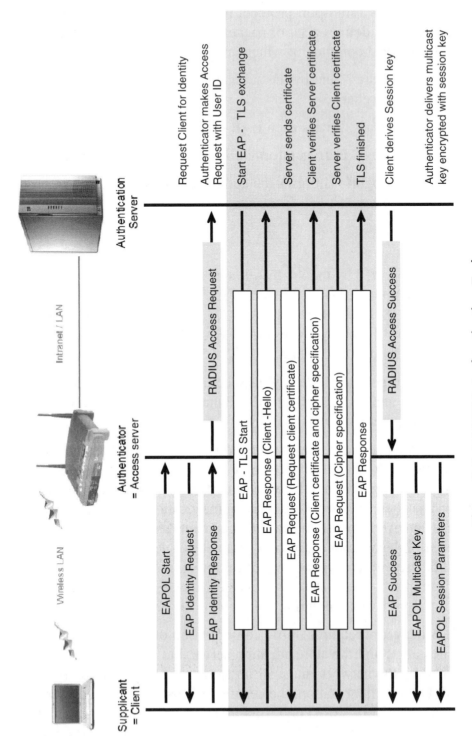

Figure 8-6: EAP-TLS Authentication Exchange

public key infrastructure (PKI) required for certificate validation can become a significant network management task.

EAP-TTLS (Tunnelled transport layer security) and PEAP (Protected extensible authentication protocol) are alternatives to EAP-TLS that dispense with the client side certificate, and the associated implementation and administration overhead. The client station confirms the identity of the authentication server by verifying its digital certificate using PKI, and then a one-way TLS tunnel is set-up allowing the client's authentication data (password, PIN, etc.) to be encapsulated as TLS messages and securely transported to the authentication server.

Public Key Infrastructure

A public key infrastructure (PKI) enables digital certificates to be used to electronically identify an individual or an organisation. A PKI requires a certificate authority (CA) that issues and verifies digital certificates, a registration authority (RA) that acts as the verifier of the CA when a new digital certificate is issued, and a certificate management system, including one or more directory services where the certificates and their public keys are stored.

When a certificate is requested, a CA simultaneously creates a public and private key using an algorithm such as RSA (see Glossary). The private key is given to the requesting party and the public key is lodged with a publicly available directory service. The private key is held securely by the requesting party and is never shared or sent across the Internet. It is used to decrypt messages that have been encrypted with the related public key, accessed from the public directory by the party sending the message.

A PKI enables a user to digitally sign a message by encrypting using a private key, and allows a recipient to check the signature by retrieving the sender's public key and using this to decrypt the message. In this way the parties can establish user authentication, as well as message privacy and integrity, without the need to exchange a shared secret.

IEEE 802.11i and WPA2

The IEEE 802.11i standard defines security enhancements for 802.11 WLANs, providing stronger encryption, authentication and key management strategies with the objective of creating a robust security network (RSN). The key features of the RSN are;

- A negotiation process that enables the appropriate confidentiality protocol for each traffic type to be selected during device association

- A key system that generates and manages two hierarchies of keys. Pairwise keys for unicast and group keys for multicast messages are established and authenticated through EAP handshakes during device association and authentication

- Two protocols to improve data confidentiality (TKIP and AES-CCMP).

Key caching and pre-authentication are also included in 802.11i to reduce the time taken for roaming wireless stations to associate or re-associate with access points.

WPA2 is the Wi-Fi Alliance's implementation of the final IEEE 802.11i standard and replaced WPA following the ratification of 802.11i in June 2004. WPA2 implements the advanced encryption standard (AES) encryption algorithm using counter mode with cipher block chaining message authentication code protocol (CCMP). TKIP and 802.11 authentication were included in the earlier release of WPA, and have been described above, while AES and CCMP are covered in the following sections.

WPA and WPA2 both support Enterprise and Personal modes, and a comparison of the main elements is shown in Table 8-3.

Table 8-3: WPA and WPA2 Compared

	Enterprise mode	*Personal mode*
WPA	Authentication: IEEE 802.1x/EAP Encryption: TKIP Integrity: MIC	Authentication: PSK Encryption: TKIP Integrity: MIC
WPA2	Authentication: IEEE 802.1x/EAP Encryption: AES-counter mode Integrity: CBC-MAC (CCMP)	Authentication: PSK Encryption: AES-counter mode Integrity: CBC-MAC (CCMP)

RSN Security Parameter Negotiation

The negotiation of security parameters between RSN capable devices is enabled by including an RSN information element (IE) which identifies the broadcasting device's RSN capabilities in beacon, probe, association and reassociation frames.

The IE identifies specific RSN capabilities; supported authentication and key management mechanisms, and ciphers for unicast or multicast messages, as described in Table 8-4.

The selection of security parameters occurs through the following exchange;

Step (1) The client station broadcasts a Probe Request.

Step (2) The access point broadcasts a Probe Response including an RSN IE.

Step (3) The client station sends an Open System Authentication request to the access point.

Step (4) The access point provides an Open System Authentication response to the client.

Step (5) The client station sends an Association Request with an RSN IE indicating its choice of RSN capabilities.

Step (6) The access point send an Association Response indicating success if the client station's selected security parameters are supported by the access point.

Notice that this exchange is unprotected and, since Open System Authentication is used (see the Section "Station Services, p. 147"), there is effectively no authentication between the client station and the access

Table 8-4: RSN Information Element content

RSN capability	*Description*
Supported authentication and key management mechanisms	RSN devices can either support 802.1x authentication and key management or 802.1x key management with no authentication
Supported ciphers	RSN devices may support any of the following ciphers for either unicast or multicast message encryption (WEP, TKIP, WRAP and AES-CCMP)

point. This does not introduce a security threat since the exchange simply serves to establish the protocol and cipher that will then be used to ensure mutual authentication and subsequent data privacy.

To provide a degree of backward compatibility with legacy equipment, networks that use RSN but do not have the required hardware to support AES will allow the use of TKIP/RC4 for encryption. This interim step towards full RSN is referred to by the term transition security network (TSN).

RSN Key Management

After security parameter negotiation, the next stage in establishing a connection between a client station and an access point is mutual authentication using 802.1x or PSK authentication, as described in the Section "802.1x Authentication Framework, p. 214". At the end of this authentication exchange, the authentication server generates a pairwise master key or alternatively, in personal mode, the key is derived from the user entered password or passphrase.

Pairwise Key Hierarchy and the 4-way Handshake

Pairwise keys are used to protect unicast messages between a client station and access point. The hierarchy of keys and the handshake that establishes and installs them are described in Table 8-5.

The purpose of the 4-way handshake, shown in Figure 8-7, is to install this key hierarchy securely in both the client station (supplicant) and the access point (authenticator). Each of the four steps involves the transmission of an EAPoL key exchange message, as follows;

Step (1) The access point (authenticator) generates a pseudo-random nonce (ANonce) and sends this to the supplicant.

Step (2) The client station (supplicant) generates a pseudo-random nonce (SNonce) and is then able to compute the PTK and derive the KCK and KEK. The supplicant sends its SNonce to the authenticator, together with the security parameters previously negotiated. The KCK is used to compute a MIC that assures the origin of the message.

Table 8-5: Pairwise Key Hierarchy

Pairwise keys	*Description*
Pairwise master key (PMK)	Starting point of the pairwise key hierarchy, either generated by the authentication server or derived from the user entered password if 802.1x authentication is not being used.
Pairwise transient key (PTK)	The PTK is derived from the PMK together with the MAC addresses of the client station and access point, and a nonce (see Glossary) provided by each party during the 4-way handshake.
EAPoL key confirmation key (KCK)	The KCK is used to assure authenticity of messages in the 4-way handshake.
EAPoL key encryption key (KEK)	The KEK is used to ensure confidentiality by encrypting messages in the 4-way and group key handshakes.
Temporal keys	The temporal key is the key that will be used in the AES cipher once the communication link has been established.

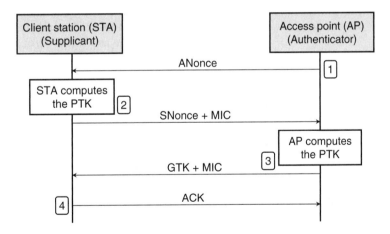

Figure 8-7: EAPoL Key Exchanges of the 4-Way Handshake

Step (3) The authenticator replies with the GTK and a sequence number, together with its security parameters and an instruction to install temporal keys. The sequence number identifies the number of the next multicast or broadcast frame, and allows the supplicant to guard against a replay attack. This message is also protected by a MIC, computed using the KCK.

Step (4) The supplicant responds confirming that temporal keys have been installed.

Group Key Hierarchy and the Group Key Handshake
Group keys are used to protect multicast or broadcast messages from an access point to all stations in its BSS. The hierarchy of keys is simpler than for pairwise keys, and consists of just two keys, as described in Table 8-6.

Table 8-6: Group Key Hierarchy

Group keys	*Description*
Group master key (GMK)	Generated by the authenticator (access point) and used as the starting point for group temporal key generation.
Group temporal keys (GTKs)	GTKs are temporal keys which will be changed periodically. For example when a station leaves the network, the GTK needs to be updated to prevent it from receiving any further broadcast or multicast messages from the access point.

The current GTK is shared with an associating client station in the third EAPoL exchange of the 4-way handshake, but a further handshake, the group key handshake, is used when the GTK needs to be updated.
A new GTK will be derived by the access point using a pseudo-random function of the GMK together with its MAC address and a nonce (GNonce). The new GTK is then distributed via the group key handshake, as follows;

Step (1) The access point sends the new GTK in encrypted unicast messages to each station in the BSS. The new GTK is encrypted using each station's unique KEK and protects the data from being tampered using a MIC.

Step (2) Each station replies to inform the access point that the new GTK is installed.

All stations will then use the new GTK to decrypt future broadcast or multicast messages.

Advanced Encryption Standard (AES)

The Advanced Encryption Standard (AES) is a cipher that was developed by Belgian cryptographers Joan Daemen and Vincent Rijmen, and was adopted as an encryption standard in November 2001 by the US National Institute of Standards and Technology (NIST) after a four year selection process. In June 2003, the US Government authorised the use of AES to protect classified information, including "Top Secret" information, provided that either 192 or 256 bit key lengths were used.

Unlike RC4, which is a stream cipher and can encrypt a message of arbitrary length, AES is a block cipher and uses a fixed message block size of 128 bits together with an encryption key of 128, 192 or 256 bits. This is a specific instance of Daemen and Rijmen's original cipher, also known as the Rijndael cipher, which can use block and key sizes of 128 to 256 bits, in steps of 32 bits.

The cipher operates on 4×4 arrays of bytes (i.e. 128-bits), and each round of the cipher consists of four steps;

Step (1) A SubByte step – where each byte in the array is substituted with its entry in a fixed 8-bit lookup table called the S-box.

Step (2) A ShiftRows step – where each row of the 4×4 array is shifted a certain number of positions, with the size of the shift differing for each row of the array.

Step (3) A MixColumns step – which mixes the four bytes in each column of the array using a linear transformation to produce a new column of four dependent output bytes. This step is omitted in the final round of the cipher.

Step (4) An AddRoundKey step – in which a second 4×4 array, called a sub key, is derived from the cipher key using a key schedule and the two 4×4 arrays are XOR'd together to generate the starting array for the next round.

The number of rounds used to encrypt each block of data depends on the key size, with 10 rounds used for 128-bit, 12 rounds for 192-bit and 14 rounds for 256-bit keys.

While AES is already a very strong cipher, AES-CCMP incorporates two additional cryptographic techniques, counter mode and a cipher block chaining message authentication code (CBC-MAC) that provide additional security between the wireless client station and the access point.

Counter Mode Operation of Block Ciphers

In the counter mode of operation of a block cipher, the encryption algorithm is not applied directly to a block of data but to an arbitrary counter. Each block of data is then encrypted in an XOR operation with the encrypted counter, as shown in Figure 8-8.

For each message, the counter is started from an arbitrary nonce and incremented according to a pattern that is known to both sender and receiver.

Counter mode contrasts with the electronic code book (ECB) (Figure 8-9) mode of operation that was commonly used with the data encryption standard (DES – the NIST predecessor of AES) in which each block of a message is encrypted with the same encryption key. With ECB, the output ciphertext has a one-to-one correspondence with the input plaintext

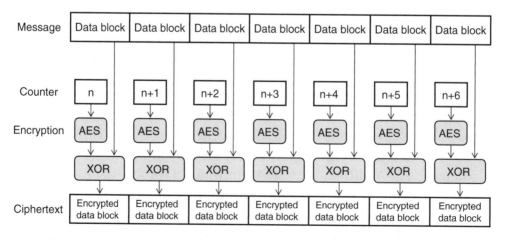

Figure 8-8: Counter Mode of Operation of a Block Cipher (AES)

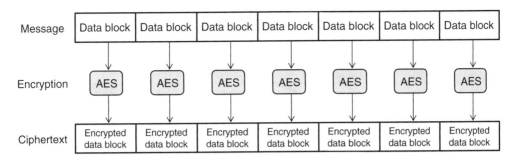

Figure 8-9: Electronic Code Book (ECB) Mode of Operation of a Block Cipher (AES)

message, so that patterns in the input data are preserved as patterns in the encrypted data, simplifying the task of discovering the encryption key. Counter mode removes this correspondence and eliminates the risk of key discovery as a result of such patterns.

Wireless robust authenticated protocol (WRAP) is a further security mechanism supported by RSN, and is based on the so-called offset codebook (OCB) mode of operation of a block cipher. This scheme has a number of advantages over AES-CCMP, including being computationally more efficient, as it performs message integrity checking, authentication and encryption in a single calculation. However, WRAP was not adopted by TGi as the base security protocol for RSN because OCB is a patented scheme, which would have introduced licensing complications into the standard.

Cipher Block Chaining Message Authentication Code (CBC-MAC)
CBC-MAC is a message authentication and integrity method that can be used with block ciphers such as AES. The acronym MIC will be used instead of MAC for the Message Authentication Code, to avoid confusion with MAC as in Media Access Control. Cipher block chaining (the CBC part) is a mode of operation of a block cipher in which the ciphertext of one block becomes part of the encryption algorithm for the next block, and so on. The CBC-MAC message authentication code is generated as shown in Figure 8-10, with the following steps;

Step (1) A 104-bit nonce is created by combining an 8-bit priority field with the 48-bit source MAC address and a 48-bit packet number.

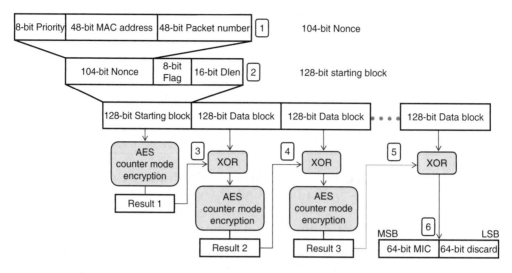

Figure 8-10: MIC Computation Using AES and CBC-MAC

Step (2) The nonce is concatenated with an 8-bit flag and a 16-bit Dlen field, which indicates the unpadded length of the plaintext data field, to generate a 128-bit starting block for block chaining,

Step (3) The starting block is encrypted using AES in counter mode (Figure 8-8) and the result is XOR'd with the first block of plaintext data from the message,

Step (4) The result is encrypted using AES in counter mode and the result is XOR'd with the second block of plaintext data,

Step (5) This chaining is repeated until the last plaintext block is XOR'd,

Step (6) The higher 64-bits of the 128-bit result are the MIC and the lower 64-bits are discarded.

Robust Security Network and AES-CCMP

The RSN security protocol, AES-CCMP (or AES counter mode-CBC-MAC protocol) defines how the three elements, AES, counter mode and CBC-MAC are used to protect data in an 802.11i implementation.

The encryption of a MAC protocol data unit (MPDU), consisting of a MAC header and a data packet, proceeds as follows (Figure 8-11);

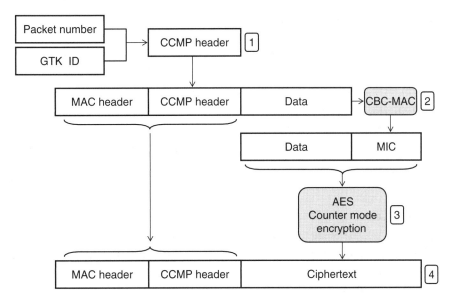

Figure 8-11: MPDU Encryption Using CCMP

Step (1) A CCMP header is constructed from a 48-bit packet number and a 3-bit GTK ID, and inserted into the MPDU, between the MAC header and the data payload.

Step (2) A CBC-MAC is computed on this extended MPDU, as described in the previous section, and the resulting MIC is appended to the plaintext data block.

Step (3) The resulting, so-called encoded data block (data + MIC) is encrypted using AES in counter mode.

Step (4) The ciphertext is appended to the unencrypted MAC and CCMP headers.

Although the MAC and CCMP headers are transmitted in plaintext, the MIC protects both the data packet and the headers from error or malicious alteration (Step (2) above).

The Sections "WEP – Wired Equivalent Privacy Encryption, p. 209", "Wi-Fi Protected Access; WPA, p. 212" and "IEEE 802.11i and WPA2, p. 219" have described the key technologies that have been developed to enable wireless LANs to achieve a high level of security. However, that security will only be assured if these technologies are supported in turn by the necessary practical measures. The following section turns to these practical security measures.

WLAN Security Measures

In order to ensure that all security vulnerabilities are recognised and addressed, every WLAN implementation needs to consider three aspects of security – management, technical and operational. A comprehensive checklist of best practice security measures in these three areas has been published by the US National Institute of Science and Technology (NIST Special Publication 800-48, see Resources the Section "General Information Sources, p. 363").

Management Security Measures

Management security measures address issues that need to be considered when designing and implementing a WLAN. Some of the key management security measures recommended as best practice in the NIST checklist are described in Table 8-7.

Table 8-7: WLAN Management Security Measures

Management security measure	*Description*
Develop a security policy for the organisation that addresses the use of wireless technology	The security policy provides the foundation for a secure WLAN and should specify the organisation's requirements including access control, password usage, encryption, control of equipment installation and administration, etc.
Perform a risk assessment to understand the value of the assets in the organisation that need protection	Understanding the value and the potential consequences of unauthorised access to the organisation's assets will provide the basis for establishing the required level of security.
Take a complete inventory of all access points and wireless devices	A physical inventory of installed devices should be cross-checked with WLAN logs as well as periodic RF sweeps for unknown devices (rogue access points).
Locate access points on the interior of buildings instead of near external walls and windows	Internal location will limit the leakage of RF transmissions beyond the required operating area and eliminate areas where eavesdropping could take place.
Place access points in secured areas	Physical security will prevent unauthorised access and manipulation of hardware.

Technical Security Measures

Technical security measures address issues that need to be considered when configuring a WLAN. Table 8-8 describes some of the key technical security measures recommended in the NIST checklist.

Changing SSID and Disabling Broadcasts

Unless disabled, the SSID is included in beacon frames transmitted by an access point about ten times every second to alert nearby stations of its presence. Every access point leaves the factory with a default SSID set, and attackers can use these IDs to access unsecured networks if they are not changed from the default values.

These default values are widely known and published on the web, and a few of the more well-known ones are "tsunami", "linksys", "wireless" and "default"! War Drivers and War Flyers report that 60–80% of identified WLANs typically still have their security setting in factory default mode, 60% have not changed default SSIDs and under 25% have enabled WEP. These figures are reported from the US, but would likely be similar in the UK.

Changing the SSID to another value is the first step in improving security. This should be done when the access point is first configured (Figure 8-12), since entering the SSID is part of the configuration process for each of the client stations that will connect to the access point. A good practice is to use an anonymous SSID that does not identify the company or organisation operating the WLAN.

After changing the default SSID, some manufacturers provide an option to disable the SSID from being broadcast in beacon frames.

This will prevent the casual eavesdropper from obtaining the SSID, but will not stop the determined hacker. The SSID is included unencrypted in the Probe Response frame transmitted by an access point when responding to a Probe Request sent by a client station attempting to connect, so a hacker equipped with "sniffer" software can extract the SSID from these messages.

The SSID is not intended to perform a security function, and changing the SSID from its default and disabling SSID broadcasts will only be effective against casual unauthorised access. Stronger measures are required to guard against a targeted attack.

Table 8-8: WLAN Technical Security Measures

Technical security measure	*Description*
Change the default SSID and disable SSID broadcast	Prevents casual access to the WLAN and requires a client station to match the SSID when attempting to associate.
Disable all nonessential management protocols on access points	Each management protocol provides a possible route of attack, so disabling unused protocols minimises the potential routes that an attacker could use.
Ensure that default shared keys are replaced by keys of at least 128-bits and periodically change keys	Manual key management will be necessary unless TKIP is installed (Section "Temporal Key Integrity Protocol, p. 212"). Best practice is to use the longest supported key length.
Deploy MAC access control lists	Access control based on MAC filtering provides additional security although, as described in the Section "MAC Addressing Filtering, p. 234", it is not secure against a technically determined attacker.
Enable user authentication and strong administrative passwords for access point management interfaces	Management control functions on access points need to be protected as well as, if not better than, the network traffic. The security policy should specify the requirement for user authentication and strong passwords.

Changing Default Shared Keys

Although the fundamental cryptographic weakness has been recognised and addressed, as was described in the Section "WEP – Wired Equivalent Privacy Encryption, p. 209", it is still good security practice to enable data encryption using wired equivalent privacy (WEP) in pre-802.11i networks, and to use the longest supported shared key length.

Some access points have factory set default encryption keys and, similar to SSIDs, it is recommended practice to change these default values when encryption is enabled. Figure 8-13 shows a typical access point configuration screen, with encryption keys being generated using an algorithm based on a password or passphrase entered on set-up.

Shared keys or passphrases must be changed to maintain security whenever a previous network user is no longer authorised to access the network or if a device that has been configured to use the network is lost or compromised.

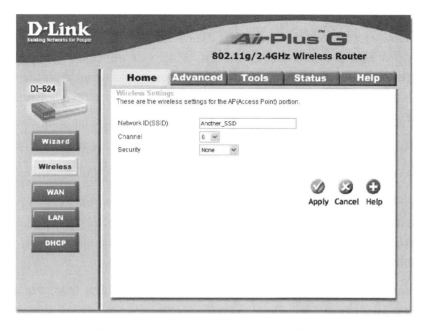

Figure 8-12: Changing SSID Default

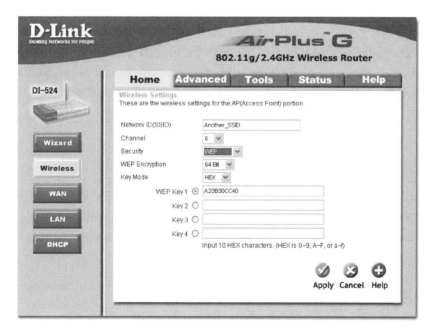

Figure 8-13: Enabling WEP and Specifying Shared Keys

MAC Address Filtering

If MAC address filtering is enabled, the access point will review every request for access against a permit list held in its memory. The permit list holds the MAC addresses of all authorised client stations, and the access point will only grant access if the requesting MAC address is found in the list.

MAC filtering appears to be a highly effective security measure but, as MAC addresses are included in transmitted data packets, a hacker could recover permitted MAC addresses by "sniffing" wireless traffic. Although the MAC address is a factory set identifier, that is, in principle, unique to an individual adapter card, a MAC address can also be temporarily reset by software to allow a device to masquerade as (or spoof) another device.

This may sound like a built-in fault, but it also has a legitimate use as some ISPs use MAC address filtering to control customer access. If it is necessary to connect more than one device to the Internet using a single ISP connection it will be necessary to reset the MAC addresses of all devices to be the same as the MAC address that was registered with the ISP.

Enabling MAC address filtering is a simple button click in the access point set-up procedure, as shown in Figure 8-14. However, keeping the access list up-to-date can be a time consuming administration task, as every authorised client station must have its MAC address manually entered into the list. In a home WLAN this will be easy enough to maintain, but in a dynamic network such as an enterprise or community WLAN, where clients may be changing fairly frequently, the cost of the administration burden may quickly outweigh the security benefits it provides.

Operational Security Measures

Operational security measures address issues that should be considered during routine operation of a WLAN. The key operational security measures recommended in the NIST checklist are summarised in Table 8-9.

Strong User Authentication

Remote authentication dial-in user service (RADIUS), as described in the Section "802.1x Authentication Framework, p. 214", is the most

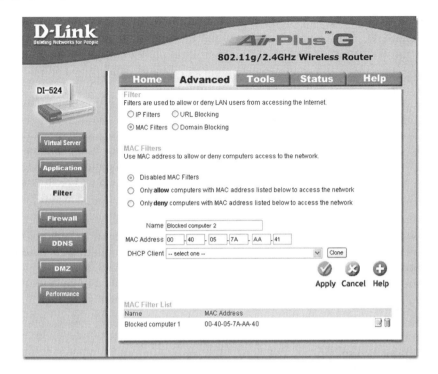

Figure 8-14: Enabling MAC Address Filtering

commonly used authentication protocol, providing centralised authentication, authorisation and accounting services.

Kerberos is another authentication protocol, developed by the Massachusetts Institute of Technology, and provides tools for authentication and strong encryption particularly for client/server applications. The source code is freely available from MIT and has also been incorporated into a range of commercial products.

Intrusion Detection

Intrusion detection software can be run to continuously monitor activity on the WLAN and generate an alarm when any unauthorised device, such as a rogue access point, is detected. Typically intrusion monitoring is based on specifying parameters that identify normal, authorised WLAN devices and traffic, as shown in Table 8-10.

When devices or network activity is identified that deviates from the specified parameters, an alarm is generated that can either be used to

Table 8-9: WLAN Operational Security Measures

Operational security measure	*Description*
Use an encrypted protocol, such as SNMP v3, for access point configuration	SNMP v3 provides encryption of access point management messages, whereas SNMP v1 and v2 do not provide the same level of security.
Consider other forms of user authentication for the wireless network such as RADIUS and Kerberos	If a risk assessment identifies unauthorised access as a key risk, authentication services or protocols, such as RADIUS and Kerberos, can provide a high degree of access security to protect confidential data.
Deploy intruder detection on the WLAN to detect unauthorised access or activity	Rogue access points or other unauthorised activity can be detected by intrusion detection software. This is a standard feature of wireless switches, described in the Section "Wireless LAN Switches or Controllers, p. 48".
When hardware is upgraded ensure that configuration settings are reset prior to disposal of old equipment	If access points are left with their secure configuration settings when they are disposed of, this sensitive information could be used to attack the network.
Enable and regularly review access point logs	Access point logs provide a basis for periodic auditing of network traffic – both authorised and unauthorised. Many intrusion detection tools can be configured to effectively perform this task automatically.

refine the parameter list or to warn the WLAN manager of an intrusion attempt and initiate countermeasures such as blocking an identified rogue device from associating or maintaining a connection with the WLAN.

Intrusion detection software can also monitor network activity to detect any attacks on the WLAN, such as DoS attacks or session highjacking, and to ensure that all authorised devices are complying with the security policies in force.

Wireless Hotspot Security

Wi-Fi hotspots provide public wireless connections to the Internet at convenient locations such as coffee shops, hotel and airport lounges. The requirement for easy public accessibility means that encryption is not

Table 8-10: Intrusion Detection Parameters

Security parameter	*Description*
Authorised RF PHY specifications	Identifies which PHY layer standards are being operated in the WLAN (e.g. 802.11a, 802.11b/g).
Authorised RF channel usage	Indicates which RF channels individual access points are configured to use.
Authorised device MAC addresses	Similar to MAC access control at the access point level, but extended to the whole WLAN.
SSID policy	Lists authorised SSIDs.
Equipment vendor	Indicates the manufacturer of the equipment that is authorised to operate on the WLAN. This provides partial MAC filtering, since the first part of the MAC address indicates the equipment manufacturer.

enabled since, prior to 802.11i, the 802.11 standards lacked the necessary key management mechanisms. As a result, responsibility for the security of information sent over the wireless connection rested firmly with the hotspot user.

With the ratification of the 802.11i standard in 2004, the next generation of hotspot services will have the full range of security measures available, including the option of EAP and RADIUS based authentication and the full 802.11i pairwise and group key management mechanisms described in the Section "Wi-Fi Protected Access; WPA, p. 212".

Until secure hotspot services become available, the technical and operational security measures described in Table 8-11 should be considered when using public hotspots.

Secure Socket Layer

Secure socket layer (SSL) is a layer 6 protocol that was developed by Netscape to protect confidential data being transmitted over the Internet, and relies on the public key infrastructure (PKI) to establish encryption keys. SSL supports several public key algorithms, including RSA (see Glossary), and several encryption ciphers, including RC4 and AES, the choice being negotiated when the secure connection is set up.

Table 8-11: Wireless Hotspot Security Measures

Hotspot Security Measure	*Description*
Set up the wireless network connection to only connect to preferred access points	Limiting automatic connection to an identified list of preferred access points will reduce the risk of connecting to an unknown access point. However, as SSIDs can be easily spoofed, this will not eliminate the risk from rogue access points.
Use VPN for connecting to a corporate network	A VPN uses an additional level of encryption to provide a protected "tunnel" through an insecure connection such as the Internet or a wireless hotspot connection.
Install a personal firewall on mobile PCs that use hotspots	A firewall acts as a barrier to prevent unauthorised access to a mobile computer through the hotspot connection. Information received from the access point will be permitted or blocked depending on configuration of the firewall. Wireless network traffic should be assigned an "untrusted" status when using a public access point.
Protect files and folders on the mobile device, using passwords or encryption. Disable file sharing.	Using the privacy mechanisms available in the PC operating system will ensure that files and data are protected even if an attacker manages to make an unauthorised connection to the mobile device.
Protect data transmitted to the access point	Confidential data, including e-mails, should be encrypted before transmission, or a secure socket layer (SSL) e-mail service can be used.
Disable the wireless NIC when not in use	Turning off the wireless NIC radio when not in use will remove a potential attack route and conserve battery power for mobile devices.
Beware of the surveillance risk in public places	Be vigilant when entering PINs or passwords in public places to ensure that confidential information is protected from surveillance.
Keep operating system and security software up-to-date	Security patches should be downloaded regularly to ensure that all software contributing to security – operating system, firewall and anti-virus software – is updated to combat all known threats.

Web sites and secure on-line services such as e-mail that use SSL security can be identified by the https:// URL prefix and by a padlock icon displayed in the browser window.

The digital certificate being used by a secure web site or service provider can be checked by double clicking the padlock icon, which will display certificate information such as the name and e-mail address of the certificate owner, certificate usage and validity date, and the web site address or e-mail address of the resource using the certificate. The certificate ID of the person or entity that certifies the information contained in the first certificate will also be provided, so that the full certification path can be checked.

VoWLAN and VoIP Security

As noted in the previous chapter, when telephony moves from the relative security of physical telephone lines and exchanges to the wireless LAN or

Table 8-12: VoWLAN/VoIP Security Measures

VoWLAN/VoIP security measure	*Description*
Separate VoWLAN phones from other LAN devices using a virtual LAN	Putting VoWLAN phones on a separate virtual LAN (VLAN) and using private, non-routable, IP addresses will prevent the VoWLAN system from being accessed or attacked via the Internet. IP phones will be needed that support the 802.1p/q VLAN standard.
Encrypt voice traffic between the phone and the PSTN gateway	Encryption can eliminate voice traffic's vulnerability to eavesdropping en route to the telephone network.
Use a firewall to protect connections between voice and data WLANs	If a connection between voice and data WLANs is required, for example, to allow desktop software to manage VoWLAN phone information, use a properly configured firewall to eliminate unwanted access routes (ports, protocols, etc.).
Use a VPN to secure VoIP traffic from remote users over the Internet	A separate user log-in to the VoWLAN virtual LAN can be used to ensure that access is limited to only the necessary ports and protocols.
Keep VoWLAN phone software up-to-date	Ensure that there is a management system in place to keep phone software up-to-date with security patches.

the unsecured Internet, additional security measures are required to secure confidential voice traffic.

Specific measures that may be taken to secure voice traffic are summarised in Table 8-12, and should be addressed in the WLAN security policy before voice services are introduced.

Summary

From the relatively simple beginnings of 802.11 security based on WEP and the quickly discovered cryptographic weaknesses of the underlying key scheduling algorithm, the pace of development and current level of sophistication of WLAN security has paralleled the advances in technology aimed at increasing speed, network capacity and other functional capabilities.

With the ratification of the 802.11i security enhancements in 2004, and the launch of the WPA2 implementation by the Wi-Fi Alliance, the foundations are in place for future WLAN installations to deliver true wired equivalent privacy over the inherently open wireless medium.

However, security is only assured when these powerful technological capabilities are supported by properly implementing the practical security measures described in this chapter.

CHAPTER **9**

Wireless LAN Troubleshooting

Analysing Wireless LAN Problems

As with any problem-solving exercise, resolving wireless LAN problems requires a systematic approach, first in analysing the symptoms to try to narrow down the possible cause, and then in investigating potential solutions.

If a step-by-step approach to network implementation was followed, as described in Chapter 7, then troubleshooting will typically be called for when a change in network or device performance occurs — either something worked before and now does not, or now does not perform as it did before.

Start by asking a number of questions to clarify the nature and extent of the problem (Table 9-1). This will help to narrow down the range of possible root causes.

The starting point for diagnosis should be a review of any recent changes in the network hardware or configuration, in the operating environment or pattern of usage, as summarised in Table 9-2.

Answering some of these questions may require further investigation, such as a repeat RF site survey if increased RF interference is a possible cause of performance degradation.

The systematic approach should continue when testing possible solutions to the problem, using the strategies described in Table 9-3.

The majority of WLAN problems fall into two categories, connectivity — when one or more client stations are unable to establish a connection to the network, and performance — when the data throughput and response time of the network does not match user expectations or previous

Table 9-1: WLAN Troubleshooting — Narrowing down the Problem

Problem identification	*Considerations*
Is it a connection problem?	A major category of problems with WLANs relate to client station connections; individual users or groups of users may be unable to connect to previously accessible network resources.
Is it a performance problem?	The second major category of problems is performance related; network coverage, speed or response time are not as expected or as previously experienced.
How extensive is the problem?	Does the problem affect just one device or is the same problem experienced by many? For example, if the problem is connectivity to the network, if only one client is affected, suspect the hardware and set-up of that device's NIC. If a whole BSS is affected, check the hardware and configuration of the access point.
How regular is the problem?	Does it occur continuously, at specific times of the day — for example around mid-day when staff are using a microwave oven in the lunch room?

experience. These two categories of problems are considered in the following sections.

Connectivity Problem Checklist

Connectivity problems can have a root cause at the PHY or MAC level, for example due to physical or configuration problems with RF hardware, or at higher levels, for example due to a failure during the user authentication process. The checklist shown in Table 9-4 can be used as a starting point for diagnosing connectivity problems.

Performance Problem Checklist

Performance problems occur in WLANs either because a transmission is not reaching the receiving station with sufficient signal to noise ratio (SNR) to be properly detected and decoded, or because an access point is overloaded and unable to cope with the volume of traffic. In turn, SNR problems can be due to low signal (coverage holes) or high noise (interference). The checklist in Table 9-5 can be used as a starting point to address performance problems.

Table 9-2: WLAN Troubleshooting — First Considerations

Recent WLAN changes	*Considerations*
Hardware changes	Have any new hardware devices been added to the network? Is the new hardware from the same manufacturer as existing hardware, or one that has been certified for interoperability?
Configuration changes	Have any configuration settings been changed recently? Operating channel changes, security mechanisms enabled, keys or passphrases changed?
Software changes	Have software or firmware upgrades recently been installed? Are any set-up changes required as a result of newly installed patches to client computer or network operating systems, device drivers or firmware?
Environmental changes — physical	Has any hardware recently been moved, such as an access point possibly creating an RF coverage hole? Have partitions walls or furniture (metal filing cabinets) been rearranged in the operating area, potentially affecting the RF propagation pattern?
Environmental changes — RF environment	Have any new wireless networks or other RF sources been installed in the operating environment or in the neighbourhood? (For example a fast-food restaurant in the next building with microwave ovens along the shared wall.)
Usage pattern changes	Have any new applications been installed which make use of the WLAN, particularly those requiring high continuous or peak bandwidth? Have there been any changes in network usage, for example a new user group with high bandwidth demands?

Troubleshooting using WLAN Analysers

Dedicated WLAN analysers are available to monitor and troubleshoot enterprise scale installations. These systems include tools for site surveying, security assessment, network performance monitoring and troubleshooting that can help the network manager in the tasks of designing, implementing, securing and finally troubleshooting WLANs. Analysers are available either as stand-alone hardware or as software packages that can be run on a laptop or handheld computer. Some examples are shown in Figure 9-1.

Table 9-3: WLAN Troubleshooting — Solution Strategies

Solution approach	*Description*
Test one hypothesis at a time	Whether it is a configuration setting or a physical layout that is being tested, make changes one at a time so that effects can be directly attributed to a single cause.
Test hardware using a known functioning substitute	The easiest way to identify a hardware fault is to replace the suspected item with a known functioning substitute — whether it is a length of CAT 5 cable, a NIC or an access point.
Keep a record	Make a note of the changes that are made, any initial settings that are altered and the resulting response of the system. This will ensure that time will not be wasted in reinvestigating old avenues and that the previous set-up can be reinstated if changes make things worse.
Check for unexpected side effects	Before declaring a problem solved, check as far as possible that new unwanted symptoms have not been introduced as a consequence of the solution of the original problem.
When everything else fails … read the instructions	Read or re-read the hardware vendor's installation instructions and check their Web site for specific information on problem diagnosis and troubleshooting.

Figure 9-1: WLAN Network Analysis Tools (Courtesy of Fluke Networks)

Table 9-4: WLAN Troubleshooting — Checklist for Connectivity Problems

Problem symptoms	*Checkpoints*
A single user is unable to connect to any access point	Check that the wireless NIC is not disabled and that the station has adequate received signal strength Check whether another client station is able to connect at the problem location Check the configuration of the client station's wireless network connection, including security settings Check that access point security mechanisms such as MAC filtering are correctly configured for the client station Replace any suspect wireless NIC with a known functioning substitute
No user is able to connect to an access point	Check the configuration of the access point, including security settings Recheck connectivity with security settings temporarily disabled Replace the suspect access point with a known functioning substitute
Users can connect to an access point but are unable to access the network	Check that the client station and access point have valid IP addresses, sub-net masks and default gateway addresses, either received from a DHCP server or manually entered Use the ping command at the OS prompt (e.g. DOS prompt) to check step-by-step connectivity, from client station to access point and from access point to a wired network computer If 802.1x authentication is in place, verify configuration and operation of the authentication server over a wired connection

Some WLAN analyser products focus on one application area, such as spectrum or protocol analysis, while others combine these specific capabilities with more general performance and security analysis tools. The typical usage of these analysis tools is summarised in Table 9-6.

With the advent of 802.11i and the increasing use of 802.1x authentication in WLANs, successful authentication becomes an additional step before a client station can successfully connect to the network. A WLAN analyser will be able to monitor each step of the EAP

Table 9-5: WLAN Troubleshooting — Checklist for Performance Problems

Root cause	*Description*
Poor SNR — low signal strength	Use a site survey tool (Section "Troubleshooting using WLAN analysers, p. 243") to test signal strength at the affected location Monitor signal strength while adjusting antenna location and orientation Consider increasing antenna gain or transmit power (up to regulatory limits) or relocating access points if signal strength remains low
Poor SNR — high noise level	Use a site survey tool to identify other 802.11 transmissions and non-802.11 interfering signals Look for and eliminate the usual suspects (microwaves, cordless phones, Bluetooth) if high noise levels are identified
Access point overload	Survey users for any changes in applications or usage patterns Turn on and review the log for the access point experiencing performance problems A high re-try count under good SNR conditions will indicate re-tries due to competing traffic Consider additional access points on non-overlapping channels, or running dual mode networks (e.g. 802.11a and g) to increase capacity

authentication process to see if a breakdown of this process is preventing user authentication and access. If the authentication server is denying a user access, the analyser results will help to determine whether the problem lies with the user's access rights or security configuration, or with the authentication server itself.

Besides a wide range of commercial tools, a number of free or open source analysis tools are also available, the most popular being NetStumbler (see the Section "Wireless LAN Resources by Standard, p. 367"). This "free for non-commercial use" tool will identify SSID, channel, encryption settings and SNR of all detected access points, and can be used to verify network configuration and to detect any rogue stations within range. In small-scale implementations it can also be used to perform a simple site survey, checking for holes in RF coverage and detecting other 802.11a/g/b networks that may cause interference.

Table 9-6: WLAN Analysers — Analysis Tools and Typical Usage

WLAN analysis tools	*Typical usage*
Site surveying	Locating non-802.11 interference sources Investigating intermittent connection problems caused by interference Monitoring of all 802.11a/b/g channels Determining a noise floor and identifying high noise or low SNR problems Verify access point channel usage and power level Mitigating channel overlap problems Performing pre-installation modelling and identifying problem areas with insufficient coverage Analysing site survey results in the light of specific technical requirements, e.g. for VoWLAN applications
Security assessment	Ensuring access points have policy compliant security configurations Visibility of encrypted network traffic (WEP, WPA, WPA2) Detection and physical location of unauthorised wireless stations Detection of wireless security attacks Identification of rejected association requests Physical location of roaming clients
Troubleshooting	Association and authentication problems Poor localised WLAN performance Lower than expected network throughput rates Changes in WLAN performance over time

Bluetooth Coexistence with 802.11 WLANs

Because the Bluetooth (802.15.1) radio shares the 2.4 GHz ISM band with 802.11b and g networks, there is a potential for RF interference between these two technologies. In fact the 802.11 FHSS and 802.15.1 specifications use the same 79 hopping channels, and the 22 MHz bandwidth of an 802.11 DSSS channel will interfere with 24 of the 79 hopping channels if adjacent channels are included.

The consequence of interference between these radios will depend on the type of spread spectrum in the 802.11 network, the transmit power of the two systems and the type of service being carried. For two interfering FHSS systems, the 802.11 system will come off worse because its hopping rate is typically 160 times slower than the Bluetooth radio.

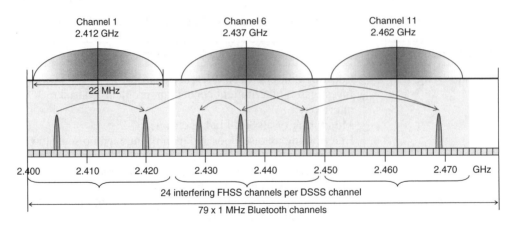

Channel 1
2.412 GHz

Channel 6
2.437 GHz

Channel 11
2.462 GHz

22 MHz

2.400   2.410   2.420   2.430   2.440   2.450   2.460   2.470   GHz

24 interfering FHSS channels per DSSS channel

79 x 1 MHz Bluetooth channels

Figure 9-2: Bluetooth and 802.11 DSSS Spectrum Overlap

This means that, in hopping over the 79 channels, the Bluetooth radio is likely to land on the same frequency as the 802.11 radio several times for each transmitted 802.11 packet. The 802.11 MAC will be issuing continued requests to repeat lost packets and network throughput will be degraded. Fortunately few 802.11 systems use the optional FHSS PHY layer specification.

The situation is a little more complex for a DSSS 802.11 system (Figure 9-2), since the direct sequence detection is inherently more robust against narrow band interference, and as the probability of a collision between a FHSS packet and a DSSS packet depends on the WLAN data packet length. In this situation the Bluetooth link is likely to be more susceptible to interference since the DSSS interference will affect 24 of the 79 hopping channels, so that some 30% of the WPAN packets could be lost. This will seriously degrade throughput, particularly for synchronous links such as voice transmission to a Bluetooth headset.

The IEEE 802.15 Task Group TG2 has developed recommended practices to reduce the interference between 802.11 and 802.15.1 radios, using two types of coexistence mechanism — collaborative and non-collaborative. Collaborative mechanisms are possible when information to minimise interference can be exchanged between the WLAN and WPAN, while non-collaborative mechanism do not require exchange of information between the two networks, but are inherently less effective. The types of

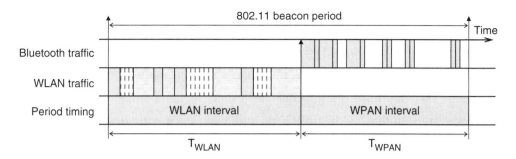

Figure 9-3: WLAN and WPAN Transmit Periods Defined in AWMA

non-collaborative approach recommended are adaptive frequency hopping, adaptive packet selection and transmit power control.

A collaborative TDMA mode termed alternating wireless medium access (AWMA) has also been recommended, where the available transmission time is divided between WLAN and WPAN transmissions, as shown in Figure 9-3. Because of the need for a communication link between the two networks, this collaborative mechanism can only operate if the two radios are located in a single host device — for example a laptop enabled for both Bluetooth and Wi-Fi.

A further collaborative mechanism is termed deterministic frequency nulling. The concept here is to reduce the narrow band interference from the 1 MHz wide FHSS signal by nulling out this frequency at the 802.11b receiver. To do this the 802.11b receiver has to follow the hopping pattern and timing of the Bluetooth transmitter, and this is achieved by embedding a Bluetooth receiver within the 802.11b receiver.

Although the Task Group is now officially in hibernation, a range of publications can be found at http://grouper.ieee.org/groups/802/15/pub. If Bluetooth interference is suspected as a cause of WLAN performance problems, an analyser such as AirMagnet's free BlueSweep utility can be used to identify active devices within the range of the WLAN operating area.

Summary of Part III

The implementation of a successful and secure wireless LAN requires attention to a broad range of issues.

Insight into the wide variety of technologies applied at the PHY and MAC layer in the different 802.11 WLAN standards provides the basis for understanding the technical capabilities of each standard and the reason behind their specific limitations, enabling the most appropriate technology to be selected for any WLAN application.

An understanding of the basic RF propagation concepts, described in Part II, provides the background to the issues that need to be considered when selecting antenna equipment and planning the physical WLAN layout for adequate RF coverage.

Appreciation of the evolving technologies applied to secure WLANs enables appropriate choices to be made in deciding on the security requirements and on-going management implications for a particular WLAN implementation.

A WLAN planned and implemented on the basis of a sound understanding of these underlying concepts and technologies will have the best possible chance of effectively meeting user performance expectations and security requirements.

PART **IV**

WIRELESS PAN IMPLEMENTATION

Introduction

A personal area network (PAN) is an interconnection of devices for personal use within the operating space of an individual — usually in the range of 1–10 metres. Wireless PANs aim to achieve this interconnectivity and give greater flexibility, mobility and freedom from the hassle of finding the right cable!

WPANs differ from WLANs in that they are not intended to replace Ethernet type local networks, giving neither the range nor, at least at present, the data capacity or variety of services of WLANs. Instead they focus on the specific information and connectivity needs of the individual — synchronising data from a desktop computer to a portable device, exchanging data between portable devices and providing Internet connectivity for portable devices.

Implementing successful WPANs is also less technically demanding than a typical wireless LAN implementation. Connections are generally quick to set up and call for little or no specific configuration.

Chapter 10 looks at the technical and practical characteristics of a range of different PAN technologies, starting with the current *de facto* standard Bluetooth and including several emerging generic and niche competitors such as Wireless USB and ZigBee. The infrared IrDA standard, which has been very widely implemented as a flexible serial port replacement, is also covered.

Implementation aspects including technology choices, security and other practical issues are covered in Chapter 11.

CHAPTER 10

Wireless PAN Standards

Introduction

Wires to connect the increasing diversity of personal devices get disconnected, broken, lost and generally get in the way of personal productivity. Getting rid of them has been the major motivation behind the efforts of the numerous working groups and other organisations involved since the late 1990s in developing standards for personal area networks.

Within the IEEE, the 802.15 Working Group was established in March 1999 with the objective of providing standards to support the interoperability of low complexity, low power devices that can be worn, carried or located in a personal operating space (POS), defined as extending 10 metres in all directions around a stationary or moving person. An overview of the wireless PAN standards developed by the 802.15 Working Group is shown in Table 10-1.

The kind of devices that were envisaged as participating in the PAN were the already ubiquitous mobile phone and then increasingly common personal stereo, pager and PDA. Considering that someone carrying a watch, mobile phone, pager, personal stereo and PDA would be equipped with two input keypads, four speakers, two microphones and potentially five LCD displays, the potential for simplification through interoperability is self-evident.

A variety of different PAN standards have been developed since the late 1990s, most notably Bluetooth and IrDA, and more recently ZigBee and Wireless USB have come on the scene. Each of these technologies has its

Table 10-1: Overview of IEEE 802.15 WPAN Standards and Task Groups

Standard	*Description*	*Application*
802.15.1	Original 2.4 GHz FHSS specification. Published in 2002.	Bluetooth
802.15.2	Recommended practices to facilitate coexistence of 802.15 wireless PANs and 802.11 wireless LANs. Published 2003.	
802.15.3a	High rate WPAN. UWB PHY with DS-UWB vs OFDM under discussion. Draft published in 2003. Overtaken by MBOA and Wireless USB. Working Group disbanded in January 2006.	
802.15.3b	MAC amendment Task Group, improving implementation and interoperability of the 802.15 MAC.	
802.15.3c	Millimetre-wave alternative PHY. 57–64 GHz unlicensed band. 1 Gbps data rate and optionally to 2 Gbps. Formed in March 2005.	
802.15.4	Low rate WPAN. DSSS 2.4 GHz, 915 and 868 MHz. Published in 2003.	ZigBee
802.15.4a	Task Group chartered to develop an alternative PHY layer. Two optional PHY specifications under consideration — a UWB impulse radio and a Chirp Spread Spectrum operating in the 2.4 unit space GHz ISM band.	
802.15.4b	Task Group chartered to address enhancements and clarifications to the 802.15.4 standard.	
802.15.5	Developing MAC and PHY mechanisms required to enable mesh networking in wireless PANs.	

own particular strengths and weaknesses in the way it addresses the challenge of delivering easy to use. In this chapter the characteristics of each of these technologies will be described, from both a technical and a practical standpoint.

Bluetooth (IEEE 802.15.1)

Origins and Main Characteristics

Research on the use of radio to link mobile phones and accessories was started by Ericsson Mobile Communications in 1994, but it was not until the Bluetooth Special Interest Group (SIG) was launched four years later, by Ericsson, IBM, Intel, Nokia and Toshiba, that the concept started to broaden beyond mobile phones to include connections between PCs and other devices.

After the IEEE 802.15 Working Group was formed in 1999 with the task of developing standards for wireless PANs, the Bluetooth SIG was the only respondent to WG15's Call for Responses, and Bluetooth and IEEE 802.15.1 soon became synonymous. Since 2000, when Bluetooth-enabled wireless headsets started to emerge, cost and power usage have reduced significantly, and Bluetooth has become a common add-on feature for many mobile phones and PDAs.

Bluetooth 1.1 is a PAN standard that operates in the 2.4 GHz ISM band at a PHY layer data rate of 1 Mbps, for an effective data rate of 721/56 kbps for asymmetric or 432 kbps for full duplex communication. Bluetooth 2.0 was ratified in November 2004, and enhanced data rate (EDR) was introduced which increased the PHY data rate from 1 to 2 or 3 Mbps.

A Bluetooth PAN can support up to eight devices in a piconet, with one device acting as master and up to seven as active slave devices. Piconets can be linked to form the so-called scatternets through the sharing of common devices, since a device can be both a master in one piconet and a slave in another. Each master device manages a piconet with a capacity of 720 kbps and a scatternet can have a much higher distributed capacity, under the control of multiple master devices.

Bluetooth supports a wide variety of different types of devices and usage models, from mobile phone headsets to PDA synchronisation. Typically different usage models will call on different parts of the Bluetooth protocol stack.

A profile is a vertical slice through the Bluetooth protocol stack (Figure 10-1), and represents the required protocols for a particular usage model. Profiles provide the basis for device interoperability and any given Bluetooth device may support a number of different usage models and therefore different profiles. Examples of the most important profiles are given in Table 10-2.

Bluetooth achieves low component cost and extended battery life by settling for limited transmission range and modest data rates. Nevertheless,

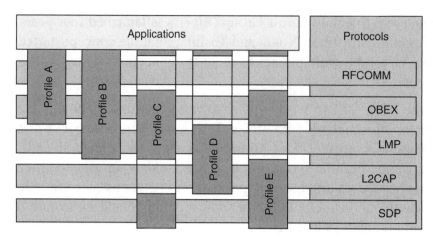

Figure 10-1: Bluetooth Application Profiles

Table 10-2: The Main Bluetooth Profiles

Profile	*Description*
Personal area networking	Enables general Internet protocol (IP) networking (including security) over an ad-hoc piconet.
Synchronisation profile	Enables the exchange of personal information such as calendar and address book data between devices.
Basic printing profile	Enables simple printing from a device to a printer. Specific printer drivers are not required in the sending device as the Bluetooth-enabled printer has the capability to decode the data sent to it to produce the required format.
File Transfer Profile	Enables a device to perform file management operations on another device's file system, including transferring, creating or deleting files or folders.
Headset profile	Enables audio data transfer between a device such as a mobile phone or a PDA and a wireless headset.
Dial-up networking profile	Enables a dial-up networking link between a PDA or other device and a remote network.
LAN access profile	Enables a device to gain access to network resources such as storage or printers by using point-to-point protocol (PPP) to connect to another device that is already in a LAN.

its effectiveness in common PAN tasks, such as telephony and short-range networking has resulted in it establishing a strong position among the available PAN options.

Protocol Stack

The Bluetooth protocol stack is illustrated in Figure 10-2. Above the Bluetooth radio (PHY layer), the Baseband, link manager protocol (LMP) and logical link control and adaptation (L2CAP) protocols correspond to the Data Link (LLC + MAC) layer of the OSI model.

The following sections describe the PHY and Data Link layers, and outline the higher level protocols up to RFCOMM and service discovery protocol (SDP). The Bluetooth stack also included protocols adopted from other sources, such as OBEX — an object exchange protocol adopted from IrDA (see the Section "IrDA Optional Protocol Stack, p. 284") and WAP (Wireless application protocol) — developed by the WAP Forum to support the delivery of Internet content over wireless links, primarily to mobile phones. Including WAP in the Bluetooth stack enables the reuse by Bluetooth profiles of software from the wireless application environment developed for WAP phones.

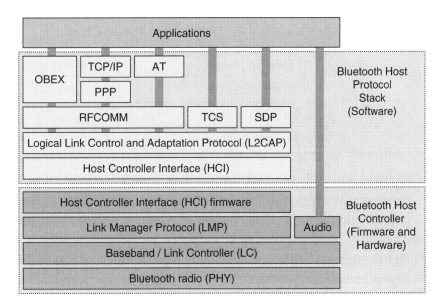

Figure 10-2: Bluetooth Protocol Stack

The Bluetooth Radio

At the physical layer, Bluetooth uses the IEEE 802.15.1 radio, which specifies a frequency hopping spread spectrum system with 1600 hops per second between the 79 channels in the 2.40–2.48 GHz ISM band, and a hopping pattern controlled by the 48-bit MAC address of the master device. In some countries, the hopping pattern is reduced to cover just 23 channels in order to comply with specific local regulations.

Gaussian frequency shift keying (GFSK) is used for the standard 1 Mbps PHY layer data rate (Bluetooth 1.2), while enhanced data rate (Bluetooth 2.0) uses $\pi/4$-DQPSK at 2 Mbps and 8-DPSK at 3 Mbps. The 2 Mbps rate is mandatory for Bluetooth 2.0 devices, while the 3 Mbps rate, relying on 8-DPSK which has a lower energy per transmitted bit and therefore a lower SNR (recall Eq. 4.1), is optional and only used over sufficiently robust links. To preserve backward compatibility with Bluetooth 1.2, GFSK is still used by Bluetooth 2.0 devices to transmit packet header information.

Three classes of RF transmitted power are defined from 0 to 20 dBm (1–100 mW), as shown in Table 10-3. Most Bluetooth devices have class 3 radios, although class 1 adapters are also available, providing a PAN range comparable with an IEEE 802.11b/g wireless LAN.

Table 10-3: Bluetooth RF Transmitter Power Classes

Class	*Max RF power*	*Range (ft)*
1	100 mW (20 dBm)	Up to 300
2	2.5 mW (4 dBm)	Up to 30
3	1.0 mW (0 dBm)	0.3–3

Transmit power control is mandatory for class 1 radios (optional for class 2 and 3), and requires transmitting devices to dynamically adjust power in order to reduce interference. This also helps to reduce power consumption and extend battery life for portable devices. To implement power control a receiver signal strength indicator (RSSI) is used to determine whether a received signal is within a defined "golden receive power range", typically between 6 and 20 dBm above the receiver sensitivity level. If the received power is outside this power range, the receiver sends a link manager protocol (LMP) instruction to the transmitter to adjust its transmit power.

Baseband Layer

The Bluetooth Baseband, sitting above the PHY layer in the Bluetooth protocol stack (Figure 10-2), manages the physical channels and links, including device discovery, link connection and management and power control.

Time division multiplexing is used to divide access to the channel across devices in the piconet. Time is divided into slots of 650 µs which are numbered according to the master device clock and allocated to links and devices by the master device.

Two types of links are supported, synchronous connection-oriented (SCO) and asynchronous connection-less (ACL). SCO links mainly carry voice transmission data and are symmetric links between the master and a single slave. To maintain the link, the master reserves transmit/receive time slots at regular intervals and, since the link is synchronous, SCO packets are not retransmitted in the event of a packet error.

ACL links connect the master to all the slave devices in the piconet (point to multi-point). The master device can establish an ACL link to any slave using time slots that are not reserved for any active SCO links. Only one ACL link can exist at a time, but the link can be to a slave that already has an SCO link to the master. For most ACL packets, packet retransmission is applied in the event of a packet error.

The Bluetooth Baseband defines 13 packet types, including 4 specifically for high quality voice and voice + data transmission. Each packet consists of a 68–72 bit access code, a 54 bit header and a payload of up to 2745 bits. The access code is used during device discovery and to gain access to a specific piconet, and the header carries the slave address as well as information for acknowledgement, numbering and error checking of packets.

The Baseband controls the process of device discovery through an inquiry procedure, which enables a device to discover other devices in range and determine their addresses and clock offsets, and a paging procedure which sets up the connection and synchronises the slave device clock to the master. Once a connection is established, a device can be in one of the

Table 10-4: Bluetooth Connection States

State	Description
Active	Devices in an active state participate in the channel. The active master schedules transmissions including regular transmissions to keep slaves synchronised. Active slaves listen for packets during ACL time slots. An active slave may sleep until the next ACL transmission if it is not addressed.
Sniff	Devices in the sniff state conserve power by listening for transmissions at a reduced rate. The inactive interval is programmable and will depend on the specific device type and application.
Hold	A data transfer can be put on hold as a power saving measure either at the request of the slave device or under the direction of the master device. In the hold state only an internal timer will be running in the slave device. The data transfer will resume immediately after the slave device returns to active mode.
Park	Devices in the park state are still synchronised but do not participate in piconet traffic. Slaves give up their 3-bit active member device address when entering this state and take on an 8-bit parked member address. Parked devices will continue occasionally to listen to transmissions in order to re-synchronise and check on other broadcast messages.

four states; active, sniff, hold, and park — in order of decreasing power consumption. Table 10-4 provides a brief description.

Higher Layer Protocols

Link Manager Protocol
The link manager protocol (LMP) is used to set up and manage Baseband connections, including link configuration, authentication and power management functions. This is achieved by exchanging protocol data units (PDUs) between the Link Managers of two paired devices. PDUs include control of pairing, authentication, initiation of sniff, hold and park modes, power increase or decrease requests and selection of preferred packet coding and size to optimise data throughput.

Host Controller Interface
The host controller interface (HCI) provides a uniform command interface to the Link Manager and Baseband layers, allowing the protocol stack to

be divided between two pieces of hardware — for example, a processor hosting higher layer software and a Bluetooth module. The host device performs upper layer functions and is able to interface with the second device performing LMP, Baseband and PHY layer functions. The two are connected through the Host Controller Transport layer, that can be either a UART, RS232 or USB interface.

Logical Link Control and Adaptation (L2CAP)

The logical link control and adaptation protocol (L2CAP) creates the logical connections between the upper layer protocols and the Baseband channels, assigning a channel identifier (CID) to each end-point of a channel. The process of establishing connections includes the exchange of information on the expected QoS between devices, and L2CAP monitors resource usage to ensure that QoS guarantees are met. L2CAP also manages the segmentation and re-assembly of data packets for higher layer protocols that use data packets larger than the Baseband's maximum transmission unit (MTU) of 341 bytes.

RFCOMM

The Bluetooth RFCOMM protocol is based on a subset of the ETSI TS 07.10 standard, and provides serial port emulation over the L2CAP protocol for cable replacement applications. RFCOMM assembles serial bit streams into bytes and data packets and provides reliable sequenced transport of the serial bit stream, using request to send/clear to send (RTS/CTS) and data terminal ready/data set ready (DTR/DSR) control signals.

One adaptation of the ETSI standard implemented in RFCOMM is a credit-based flow control mechanism that limits frame transmission rate to ensure that the receiving device's input buffer does not overflow. If the credit counter for a connection reaches zero, RFCOMM will stop and wait until it receives more credit from the receiving device, indicating that the input buffer is able to receive data.

SDP

The service discovery protocol (SDP) enables applications to discover what services are available on devices in the piconet, and to determine the characteristics of available services.

Services are discovered using a request/response model, where an application sends out a protocol data unit requesting information on services available on a particular L2CAP connection, and awaits a response from the target device.

Service discovery can be either by searching, requesting information on a specific desired service, or by browsing, requesting information on all available services.

Bluetooth in Practice

A Bluetooth piconet is established by the process of device discovery and pairing between master and discovered slave devices. This process can be repeated many times to create a PAN with up to 7 active slave devices, although 255 slave devices can remain connected to the piconet in the parked state.

During the pairing process, the slave device receives a frequency hop synchronisation data packet, based on the master device's 48-bit MAC address, in order to follow the frequency hopping pattern. Once this low-level connection is made, the master device establishes a service discovery protocol (SDP) connection to determine which profile will be used to communicate with the slave device. LMP is then used to configure the link according to the specific service requirements.

The two devices also exchange a passphrase which may be used to generate an encryption key to ensure secure communication, depending on security settings which are discussed further in Chapter 11.

An example of a Bluetooth piconet, and the associated profiles used for the various device pairings is shown in Table 10-5.

If the mobile phone in the above example is then paired with a Bluetooth enabled headset, it will become the master device in a second piconet, thus forming a scatternet. The mobile phone will then time-share between the two piconets. Time slots will be allocated to it by the master device in the first piconet and it will in turn allocate time slots to the headset and any other paired devices in the second piconet.

Since the frequency hopping patterns of the two piconets are determined by different master device MAC addresses, they are not coordinated and there will be random data packet collisions when the same frequency is

Table 10-5: Example of a Bluetooth PAN (Piconet) and Associated Profiles

Device	Master/Slave	Profile
Laptop computer	Master	
Printer	Slave	Serial port profile
PDA	Slave	Synchronisation profile
Laptop computer	Slave	File transfer profile
Mobile phone	Slave	Dial-up networking profile

chosen. However, this will occur very infrequently (statistically once in every $79 \times 79 = 6241$ data packets) and so will not significantly affect data throughput.

Bluetooth Usage Examples
To give an idea of the way Bluetooth is used in practice it is useful to look at the steps involved in setting up a number of connections representing different profile usage.

Synchronising a PDA over Bluetooth   The steps in setting up this service, which uses the serial port and synchronisation profiles, are shown in Table 10-6.

Table 10-6: Usage Model — PDA Synchronisation

Step	Description
1	Pair the two devices (PDA and desktop or laptop), exchanging PIN/passkey.
2	Use installed Bluetooth software on the host desktop/laptop to associate the incoming Bluetooth serial port with a specific COM port (e.g. COM3). Security of the link (encryption) will usually be specified in this step, and automatic synchronisation on connection may also be specified.
3	Use synchronisation software on host desktop/laptop (e.g. Microsoft ActiveSync) to identify the serial port for synchronisation with the COM port specified in Step (2).
4	Establish a connection with the host desktop/laptop from the PDA.
5	Initiate synchronisation, unless automatic synchronisation on connection was selected at Step (2).

Internet connection from a PDA over Bluetooth The steps in setting up this service, which uses the LAN access profile, are shown in Table 10-7.

Table 10-7: Usage Model — Internet Connection from a PDA

Step	*Description*
1	Pair the two devices (PDA and desktop or laptop), exchanging PIN/passkey.
2	Use installed Bluetooth software on the host desktop/laptop to configure network access in order to allow other devices to connect to the Internet/LAN through the host.
3	Use the connection wizard or similar software on the PDA to establish a connection to the Internet through the host computer (rather than via a separate dial-up device such as a mobile phone).
4	When the PDA attempts to connect to the host computer, authorise the connection from the host.
5	If required, enter dial-up and/or log-on information to complete the connection.

Dial-up Networking over a Bluetooth Enabled Mobile Phone The steps in setting up this service, which uses the dial-up networking profile, are shown in Table 10-8.

Table 10-8: Usage Model — Dial-up Networking via a Bluetooth Enabled Mobile Phone

Step	*Description*
1	Pair the two devices (the mobile phone and the computer requiring dial-up access), exchanging PIN/passkey.
2	Use installed Bluetooth software on the computer to confirm via dial-up networking properties that a Bluetooth modem is installed for the mobile phone. If not this can be downloaded from the phone maker's Web site.
3	On the computer, connect to dial-up networking (e.g. in Windows, double click on the dial-up networking icon in the Bluetooth devices folder).
4	Enter username, password and ISP dial-up number when requested. Dial-up, user authentication and registration will follow.

Application software such as dial-up networking or Internet connection may offer the option to save username and password information on the mobile device. This can be a security risk if the mobile device is lost and is not protected by a password. Bluetooth security, including a checklist of operational security measures, is discussed in the next chapter.

Current and Future Developments

Bluetooth's FHSS physical layer specification provides limited scope to compete with the higher data rates that will be available using wireless USB and other UWB radio based PAN standards. An increase in data rate to 3 Mbps was achieved in Bluetooth 2.0, but much higher data rates are expected to be in demand by users with the ever growing volume of digital content being transferred between devices.

In May 2005, the Bluetooth SIG announced that it is working with ultra wideband developers to extend Bluetooth capabilities to achieve the data rates that will be required by high speed applications such as delivering digital video to portable devices in a PAN. This development clearly aims to leverage the strong Bluetooth brand which has considerable market strength although, from the user's point of view, backward compatibility will be expensive to achieve as UWB and the current 2.4 GHz FHSS PHY are not interoperable.

Wireless USB

Origins and Main Characteristics

Wireless USB is the result of the drive by the USB Implementers Forum to ensure that the highly successful wired USB interface evolves into the wireless future. The Wireless USB Promoter Group was formed in February 2004 to create a wireless extension to USB that would apply the original USB principles of ease-of-use, compatibility and low cost, to high speed wireless technology. The strong industry support and brand recognition of USB are assets that the group hopes will give wireless USB (WUSB) a head start in the wireless PAN sector. Retaining a strong link with wired USB is central to the design goals of wireless USB (Table 10-9).

Wireless USB uses ultra wideband (UWB) radio technology to deliver a PHY layer data rate of 480 Mbps, with low power consumption and a range of up to 10 metres. This will enable wireless USB to comfortably

Table 10-9: Wireless USB Design Objectives

Design goal	Description
Preserve USB software infrastructure	Wireless USB is designed to use the same software interface and device drivers as USB.
Preserve the smart host simple device model	As with USB devices, wireless USB keeps the device simple and leaves management of network complexity to the host.
Enable power efficiency	Many devices that took power via a wired USB connection will become battery powered under wireless USB. Effective power management mechanisms will be required.
Provide wired equivalent security	Device authentication and data encryption aim to provide the same level of security as wired USB.
Ease of use	Plug and play is the user expectation for USB devices, and wireless USB is designed to continue that tradition.
Preserve USB investment	Wireless USB defines a specific device class (Wired Adapter) to allow wired USB devices or hosts to support wireless USB devices.

stream video to multimedia consumer electronic devices, as well as offer high-speed connections to PC peripherals and other mobile devices.

Revision 1.0 of the wireless USB specification was released in May 2005. Wireless USB devices share bandwidth through a host-scheduled TDMA based media access protocol. A hub and spokes logical topology is used, with each host supporting up to 127 devices and, like wired USB, system software is designed to accommodate devices connecting to or disconnecting from the host at any time. A "dual role" device is also defined to enable peer-to-peer connections.

Protocol Stack

The foundations of the wireless USB protocol stack are the PHY and MAC layer specifications developed by the MBOA–Special Interest Group (SIG). Since the merging of the MBOA–SIG and the WiMedia Alliance in March 2005, these specifications are now being finalised and promoted by the combined WiMedia–MBOA Alliance.

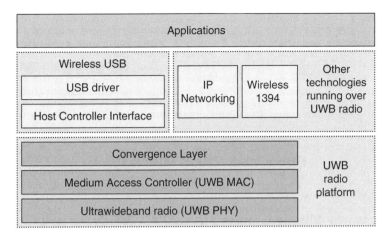

Figure 10-3: Wireless USB Protocol Stack

Wireless USB is one of the number of higher level technologies that will run over the MBOA PHY and MAC (Figure 10-3) layers, and the wireless USB specification defines the way in which a wireless USB communication channel is established using these lower layers. At the application level, wireless USB is functionally identical to USB 2.0, except for some enhancements to the isochronous data communication model to allow for the relative unreliability of the wireless PHY layer compared to a wired USB connection.

Wireless USB Radio

The wireless USB PHY layer is the Multiband OFDM Alliance (MBOA) UWB radio, operating across the 3.1 to 10.6 GHz frequency bands. Support for data rates of 53.3, 106.7 and 200 Mbps is mandatory for wireless USB devices, with additional rates up to 480 Mbps being optional for devices and mandatory for hosts.

Support for bands 1 through 3 (Channel 1 — see Figure 10-4) is mandatory for all wireless USB implementations, with optional support for other band groups. All time-frequency codes (TFC) for each band group supported must also be supported (see the Section "Multiband UWB, p. 122").

Media Access Control Layer

The WiMedia MAC has been specifically developed to address the shortcomings of previous MACs, such as pre-802.11e Wi-Fi and

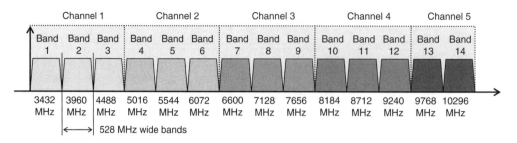

Figure 10-4: WUSB MBOA Bands

Bluetooth, in providing guaranteed quality of service for real time video and audio streaming applications, as well as robustness to changing network topology. The key design features of the WiMedia MAC are summarised in Table 10-10.

MAC layer timing is defined within the superframe structure shown in Figure 10-5. Each 65 ms superframe is divided into 256 media access slots (MAS) each of 256 μs duration. The leading MASs in each superframe are used as a beacon period, during which devices exchange information with the host on their capabilities and resource requirements. Devices can reserve one or more medium access slots using distributed reservation protocol (DRP) messages during the beacon period. This enables applications to guarantee media access for isochronous data streams.

Table 10-10: WiMedia MAC Key Design Features

Design feature	*Description*
Distributed network control	At the MAC level, responsibility for medium control is shared by all devices, reducing vulnerability to single-point failure and eliminating the bandwidth penalty of maintaining central control.
Prioritised access mechanisms	A TDMA system allows devices either to reserve guaranteed medium access slots, or contend for access during a prioritised contention period.
Network management efficiency	The bandwidth overhead associated with the MAC protocol scales with the number of devices, assuring a low overhead for networks with few devices.

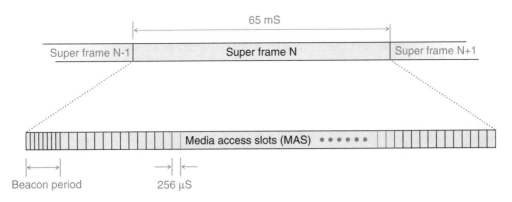

Figure 10-5: WiMedia MAC Superframe Structure

Media access slots that are not reserved are available for use by any device, based on prioritised contention access (PCA). Here the prioritisation mechanism ensures that asynchronous but time-sensitive transmissions, such as those to and from a user interface device, will get priority over other non-time-sensitive devices. The contention access period is also used to increase the robustness of isochronous connections by providing an opportunity for the MAC layer to retransmit any failed packets from the DRP period.

The Wireless USB Channel

The wireless USB specification defines the way in which a wireless USB channel is established within the superframe structure and associated MAS reservation and control mechanisms.

A host creates the wireless USB channel using DRP to reserve media access time slots that will be used for communication by all devices in the cluster. The host controls the channel using a sequence of control packets called micro-scheduled management commands (MMCs), which are transmitted during the reserved media access slots. These commands are used to dynamically schedule and control channel time for communication between the host and devices in the cluster.

The MMC is a broadcast packet that contains a cluster ID to enable devices to identify control packets for their cluster. Each MMC specifies the breakdown into micro-scheduled channel time allocations (MS-CTAs) of reserved time until the next MMC. These allocations are used for data communication within the cluster, with the direction and use of each MS-CTA being specified in the preceding MMC.

As shown in Figure 10-6, although the MAC layer media access slots which support the channel will be discontinuous in time, the MMCs effectively pull these slots together into a contiguous channel for communication within the cluster.

To ensure strict compliance with the MAC layer requirements, such as beaconing and distributed control, a wireless USB host is required to implement the full WiMedia MAC protocol. Other devices are only required to implement the wireless USB protocol that operates within the wireless USB channel. Wireless USB defines mechanisms to ensure that a host is aware of and respects the DRP reservations of any possible hidden neighbours of devices in its cluster. Three device classes are defined, as shown in Table 10-11, with different levels of "awareness" of the full MAC protocol.

Directed beaconing devices have the capability to capture control protocol information transmitted by devices that are not part of the cluster and to transmit this back to the host via the wireless USB channel. This enables the host to periodically check for the presence of devices complying with the WiMedia MAC that are outside its range, and to ensure that MAC

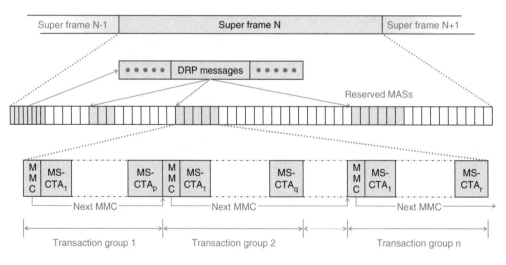

Figure 10-6: Wireless USB Channel within MAC Layer Superframes

Table 10-11: Wireless USB Device Classes

Device class	*Description*
Self-beaconing devices	These devices comply fully with MAC level protocols and do all related beaconing.
Directed beaconing devices	These devices perform beaconing and other MAC functions under the direction of the wireless USB host device.
Non-beaconing devices	These devices have reduced transmit power and receiver sensitivity and can only operate in close proximity to the host.

reservations made by the host and by non-cluster devices are mutually observable and respected.

Directed beaconing devices must support three functions to enable these capabilities;

- A Count Packets function; by periodically counting packets during the beacon period, the host can determine whether a directed beaconing device has any hidden neighbours.

- A Capture Packets function; by capturing the beacon transmission of a hidden neighbour, the host can determine its DRP reservations, and adjust its own reservations if necessary.

- A Transmit Packet function; by providing the appropriate beacon data and instructing a directed beaconing device on when to transmit it, the host can inform the hidden neighbour of the presence and DRP reservations of nearby devices in the cluster.

These control mechanisms also ensure that several wireless USB clusters can spatially overlap with minimum interference.

The wireless USB protocol defines the packet formats and controls the various types of data transfers (bulk, isochronous and control) within the micro-scheduled wireless USB channel. The protocol also provides control of a range of measures designed to mitigate the effect or RF interference on transfer reliability, including control of transmit power, bit rate and size of the data payload within a transmitted packet, as well as bandwidth and RF channel switching.

Wireless USB in Practice

One of the design goals of wireless USB is that it should preserve the "plug and play" ease of use associated with wired USB. The equivalent wireless concept of "turn on and use" will require wireless USB devices to automatically install drivers and security features when turned on for the first time, as well as identifying and associating with other devices with a minimum of user input for authentication.

Besides the hub and spokes topology that allows a host to control up to 127 end-point devices, wireless USB allows a dual role device (DRD) to function as both a host and a device simultaneously. This allows simple peer-to-peer connections between two dual role devices, with each device in its host role managing a separate wireless USB channel (called the default and reverse links) over a common MAC layer channel.

DRDs can also be connected to one or more wireless USB channels as a device while at the same time providing a wireless USB channel for other devices as a host. An example of this combination scenario would be a wireless USB printer acting as a device to a laptop computer and as a host to a digital camera.

Although wireless USB devices are still under development, several features of the specification, particularly of the MAC layer, point to some important characteristics that will have practical implications.

The WiMedia PHY and MAC have features that allow the distance between devices to be determined by measurement of the two-way transfer time (TWTT) of a message between devices. Simple MAC to MAC transactions will allow devices to exchange measurement frames and pass a distance calculation up to the application layer. With several devices distributed in 3D, triangulation can then be used to establish the spatial location of each device. This can be put to a variety of uses, from the trivial (helping to find a lost PDA) to more significant location specific services.

Current and Future Developments

Although wireless USB is still in its infancy, it seems likely that, building on the foundation established by wired USB and the MBOA UWB radio platform (which is likely to be common to several other technologies including W1394), wireless USB will quickly become established as a standard for wireless interconnection of PC peripherals and other consumer electronic devices.

The wireless USB architecture and protocol are scalable to higher data rates and, as the MBOA UWB radio platform evolves, data rates of 1 Gbps and beyond are likely to be achieved.

ZigBee (IEEE 802.15.4)

Origins and Main Characteristics

ZigBee is a standards based technology addressing the needs of remote monitoring and control, and sensory networks. The ZigBee Alliance was formed in November 2002 with the aim of exploiting the IEEE 802.15.4 radio, and the ZigBee 1.0 specification was finalised in March 2005. ZigBee delivers low data rate communications, up to 250 kbps, with ultra-low power consumption, and aims to provide a device control channel, rather than the high rate data flow channel targeted by technologies such as Wireless USB.

Some of the specific steps that have been taken to achieve ultra-low power consumption are;

- reducing the amount of data that is transmitted including the frame overhead (addressing and other header information)

- reducing the transceiver's duty cycle including power management mechanisms for power-down and sleep modes

- targeting a limited operating range of around 30 metres.

As a result, a ZigBee network will typically require only 1% of the power of an equivalent Bluetooth PAN, resulting in a battery life of months to years.

Two device classes will be defined; full function devices which implement the full protocol stack and are capable of being co-ordinating nodes and connecting with any type of device in any topology, and reduced function devices which will implement a simplified protocol set and will be limited to acting as end-nodes in simple connection topologies (star or peer-to-peer).

Every ZigBee network has a unique full function personal area network co-ordinator (similar to a Bluetooth Master device) which is responsible for network management tasks such as associating new devices and beacon transmissions. In a star network all devices communicate with the PAN co-ordinator, while in a peer-to-peer network individual full function devices are able to communicate with each other.

The very low target cost ($1–$5) will make ZigBee ideal for wireless monitoring and control applications such as residential and commercial building automation (smart home) and industrial process control. In the home ZigBee can enable the creation of a home area network (HAN), allowing the many devices currently controlled by a proliferation of non-interoperable remote controllers to be brought under the command of a single control unit.

Protocol Stack

The ZigBee 1.0 specification includes an upper layer protocol stack (Figure 10-7) building on the IEEE 802.15.4 PHY and MAC layer specifications that were finalised in May 2003. Logical network control, security and applications layer are optimised for time-critical applications, with very fast device wake-up and network association times, typically in the region of 15 and 30 ms respectively.

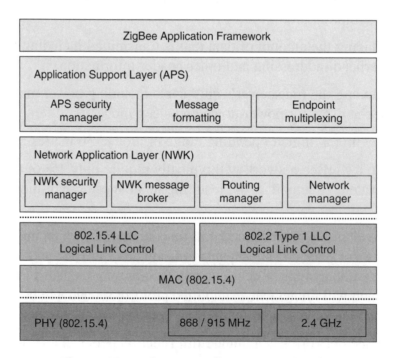

Figure 10-7: ZigBee Outline Protocol Stack

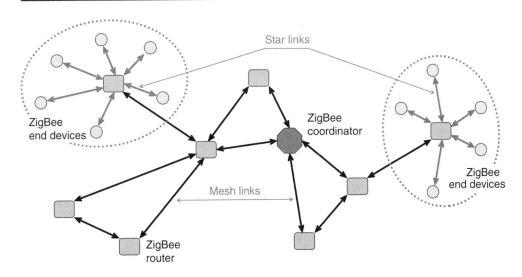

Figure 10-8: ZigBee Supported Topologies

The network layer is responsible for the usual tasks of network start-up, associating, dissociating, address assignment to devices, security, frame routing, etc. Multiple network topologies are supported by the network layer, as shown in Figure 10-8. The mesh topology can extend a network to up to 64,000 nodes through the use of ZigBee routers, which achieve efficient routing using a request–response algorithm rather than the use of router tables.

The general operating framework (GOF) is an integration layer linking the application and network layers, and maintains an overview of device descriptions and addresses, events, data formats and other information that is used by the applications to command and respond to networked devices.

Finally, at the top of the stack and similar to Bluetooth, application profiles are defined to support specific usage modes. For example, a lighting profile would include sensors for light level and occupancy as well as the switching and dimming of load controllers.

The ZigBee PHY layer

The IEEE 802.15.4 specification is the basis of the ZigBee PHY layer, and uses a number of RF bands and data rates as shown in Table 10-12.

Table 10-12: IEEE 802.15.4 Radio Frequencies and Data Rates

Band	Coverage	Channels	Data rate
2.4 GHz	Worldwide	16 channels	250 kbps
915 MHz	America	10 channels	40 kbps
868 MHz	Europe	1 channel	20 kbps

The availability of 16 non-overlapping channels in the 2.4 GHz ISM band allows for 16 simultaneous operating PANs.

In the 2.4 GHz band, direct sequence spread spectrum is used, with one of the 16, 32-bit chipping codes mapped onto a 4-bit data symbol. The chipped data stream is modulated onto the carrier using offset-QPSK at a transmission rate of 2 Mcps (million chips/sec). This chip rate translates into a raw data rate of 244 kbps.

The 802.15.4 radio specifies a transmitter power of at least −3 dBm (0.5 mW) and a receiver sensitivity of −85 dBm for the 2.4 GHz band or −91 dBm for the 915/868 MHz bands, giving an effective range for a ZigBee network in the region of 10–70 metres, depending on transmitter power and environmental conditions.

Media Access and Link Control Layer

ZigBee will use the IEEE 802.15.4 MAC with 15.4a modifications, to support up to 64,000 nodes in a variety of simple connection topologies. In an extended network, device access to the physical channel is controlled using a combination of TDMA and CSMA/CA.

A "superframe" structure (Figure 10-9) breaks the time period between beacon transmissions into 16 time slots. Beacons are transmitted by the PAN co-ordinator at predefined intervals of between 15 ms and 252 seconds, and are used to identify and synchronise devices in the network. These beacon messages are sufficiently infrequent not to suffer collisions, and are therefore not subject to CSMA/CA.

The 16 time slots are divided into two access periods, a contention based access period when devices will use CSMA/CA to determine periods when they are able to transmit, and a contention free period when devices will use guaranteed time slots, assigned by the PAN co-ordinator (TDMA). The

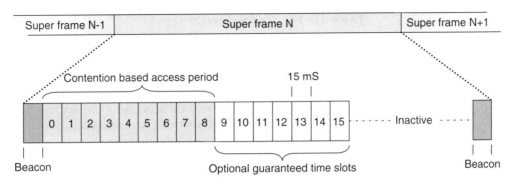

Figure 10-9: ZigBee Superframe Structure

combination of predefined beacon intervals and guaranteed transmission time slots allow sensing devices to conserve power during extended sleep periods, waking only to check in on beacons or to use a guaranteed transmission slot.

CSMA/CA used during the contention access period is very similar to that used in IEEE 802.11 networks. Devices listen before transmitting and backoff a random number of time slots before attempting to transmit. The backoff period is doubled each time a collision is sensed. A transmitting device can set a header bit to request an acknowledgement (ACK) for a message and will retransmit a message a fixed number of times if the ACK is not received.

A non-beacon mode is also defined for application in which the controller does not need to conserve power and where device transmission is so infrequent that a collision avoidance strategy is an unnecessary overhead. An example would be a security system, with a mains powered controller and (hopefully) very infrequent security alerts.

Three device types are recognised in the IEEE 802.15.4 MAC specification; full function devices, the network co-ordinator that is a special type of full function device, and reduced function devices. These device types are described in Table 10-13.

Devices use either full 64-bit or short 16-bit addresses, and transmitted frames can include both destination and source addresses. This is necessary for peer-to-peer connections and is also important in mesh topologies, to provide robustness to single point failures in the network.

Table 10-13: ZigBee Device Types

Device type	Description
Full function device (FFD)	FFDs carry all features specified by the IEEE 802.15.4 standard. They have additional memory and computing power to perform network routing functions and can be used as edge devices where the network interacts with external devices.
Network co-ordinator	The PAN co-ordinator is the most complex of the device types with the largest memory and computing power. It is a full function device that maintains overall control of the network.
Reduced function device (RFD)	RFDs have limited functionality in order to reduce complexity and cost. They can only communicate with FFDs and will generally be used as network edge devices.

ZigBee in Practice

ZigBee networks will cover a variety of short-range, low data rate applications from PC peripheral interfacing to industrial control, as shown in Table 10-14.

A typical ZigBee home automation network might cover the first three of these application areas, bringing together control of lighting, security, home entertainment and a variety of PC peripherals into a single network.

Table 10-14: ZigBee Application Areas

Application area	Application examples
PC peripherals	Mouse, keyboard and joystick interfaces
Consumer electronics	Home entertainment system (TV, VCR, DVD, audio system) remote control
Residential and other building automation	Security and access control, lighting, heating, ventilation and air conditioning (HVAC), irrigation
Healthcare	Patient monitoring, fitness monitoring
Industrial control	Asset management, industrial process control, energy management.

Although a ZigBee network may be competing for access to the 2.4 GHz ISM band with Wi-Fi and Bluetooth networks, as well as a range of other control and communication devices, ZigBee is likely to be very robust to potential interference since the duty cycle of a ZigBee device will generally be very low. The CSMA/CA mechanism, together with backoff and retry if no acknowledgement is received, means that if there is interference, ZigBee devices can simply wait for an opening and keep trying until packet reception is confirmed. The low duty cycle and low data volumes also mean that ZigBee devices are unlikely to cause significant interference to overlapping Wi-Fi or Bluetooth networks.

Mesh Implementation Considerations
A number of special considerations arise in order to ensure the successful implementation of a functional, robust and reliable ZigBee mesh network, for example in an extensive building or industrial automation application.

Mesh functionality requires that every device is able to communicate with at least one, and preferably several other devices, providing one or more paths to the central controller or point of exit from the mesh. Clearly this is also the case for WLAN installations, but in a ZigBee mesh, where devices may be installed in or concealed by machinery or pipework, special attention to signal strength site surveying will be required.

Weak or broken links in the mesh can result in the separation from the main mesh of one (orphan) or a group of devices (split mesh). The robustness of a mesh will also be adversely affected by any susceptibility to single link failures. During planning, the mesh topology should be carefully reviewed for single point vulnerability and after installation the signal strength of each link should be checked to ensure that it is sufficient, particularly at aggregation points where data throughput is expected to be high.

In extensive installations, with many tens or hundreds of installed devices, maintaining good records of device location and service history will also be important to ensure that the reliability of the network can be efficiently maintained.

Current and Future Developments
ZigBee is one of the number of competing technologies in the area of sensor networking and remote control. ZigBee's strengths are the IEEE

Table 10-15: Sensor Networking Technologies — Specific Strengths

Technology/Alliance	*Technology strengths and key features*
Insteon	Combination of 132 kHz power line modulation and wireless networking (915 MHz ISM band). Backward compatible with X-10 home automation systems.
Dust Networks	Full mesh networking — every device has message routing capability. Proprietary 25 channel FHSS radio in the 915 MHz ISM band.
Z-Wave	Full mesh networking. Proprietary 868/915 MHz ISM band radio (9.6 kbps, BFSK modulation) as well as 802.15.4 compliant products.

standard on which it is based and the broad industry alliance which will assure interoperability over a wide product range.

Competing sensor networking technologies, such as those offered by Dust Networks, Millennial Net and Insteon are proprietary, although Z-Wave is also backed by an industry alliance and aims to develop IEEE 802.15.4 based products. Each of these proprietary technologies has its own specific advantages, some of which are shown in Table 10-15.

Future developments from the ZigBee Alliance could include a ZigBee 2.0 specification based on the enhanced low data rate specification currently under development by the IEEE 802.15 TG4a. This task group is working on an alternate physical layer specification for the 802.15.4 standard which aims to deliver a location capability with an accuracy of 1 metre or better, higher data throughput, ultra-low power and longer range, as well as lower cost.

IrDA

Origins and Main Characteristics
The Infrared Data Association (IrDA) began life in 1993 as a non-profit organisation with the aim of promoting the use of infrared communication links between PCs and other devices by developing and supporting standards to ensure hardware and software interoperability. The group

published its first standard in June 1994, including the specification of the Serial Ir Link (SIR) which uses infrared to replace the serial interface cable. Since then IrDA has grown to be the most widespread wireless connection technology, with over 250 million IrDA compliant interfaces shipped in 2004.

IrDA is a low-cost, low-power, serial data connection standard that supports a half-duplex, point-to-point connections with a range of at least one metre and a data rate of up to 115 kbps (SIR and standard power mode). IrDA operates at a wavelength of roughly 1 µm compared to 12.5 cm for Bluetooth at 2.4 GHz.

Unlike the omnidirectional coverage achievable with RF transmitters, IrDA's point-to-point connection model requires the Ir transceivers to be aligned within ±30° in order for the receiver to be illuminated with the required minimum power density (Figure 10-10). This physical requirement makes IrDA well suited for some applications, like secure simple object exchange, but not so well suited for others, such as ad-hoc networking or supporting audio or telephony headsets.

IrDA has also been successful in developing generic protocols, such as the OBEX (object exchange) protocol, which allows devices to exchange objects such as business cards, files, pictures and calendar items. This was introduced by IrDA in 1997, and has been broadly adopted as a simple solution for object exchange over various transport options, including TCP/IP and Bluetooth.

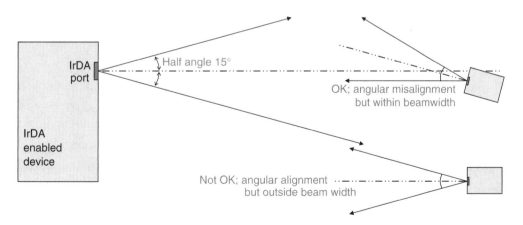

Figure 10-10: IrDA Device Alignment

Protocol Stack

The IrDA protocol stack supports data link initialisation and shutdown, connection start-up and disconnection, device address discovery and conflict resolution, data rate negotiation and information exchange.

On top of the PHY layer specification, IrDA has two mandatory protocols at the MAC level, as well as a number of optional layers that are available for specific usage models. The mandatory protocols are the link access (IrLAP) and link management (IrLMP) protocols (Figure 10-11).

IrDA PHY Layer

The IrDA infrared physical specification (IrPHY) covers aspects of the infrared beam such as wavelength, minimum and maximum power levels or irradiance in mW/sr (milliwatts per steradian) and beam angle, as well as the physical configuration of the optical components. Infrared wavelengths of 0.85–0.90 μm are specified, since light emitting diodes and optical detectors for these wavelengths are readily available and at low-cost. Two power modes are specified — standard and low-power. Link distances of up to 0.2 m are possible in low-power mode, with a maximum power intensity of 28.2 mW/sr or up to 1m in standard mode with 500 mW/sr maximum power intensity.

The IrPHY specification also defines the encoding and framing of data for various transmission speeds, as summarised in Table 10-16.

SIR is an asynchronous format that uses the same data format as the standard UART (1 start bit, 8 data bits, 1 stop bit). An RZI modulation

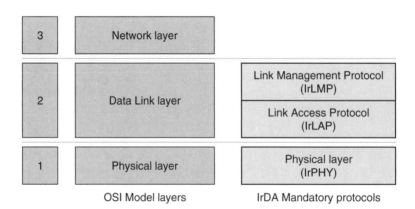

Figure 10-11: IrDA Mandatory Protocol Stack and the OSI Model

Table 10-16: IrDA Data Rates and Modulation Methods

Transmission type	*Data rate*	*Modulation*
SIR (Serial Ir)	9.6–115.2 kbps	RZI
FIR (Fast Ir)	0.576–1.152 Mbps	RZI
	4 Mbps	4PPM
VFIR (Very Fast Ir)	16 Mbps	HHH

method is used (Figure 10-12), with a short Ir pulse being transmitted for each zero data bit. The Ir pulse is shortened to nominally 3/16th of the bit duration in order to reduce the LED power consumption.

FIR and VFIR are synchronous transmission formats using RZI, 4-PPM or HHH modulation methods. The Ir pulse durations are nominally 25% of the bit or symbol duration.

All transmissions start at the lowest data speed of 9.6 kbps in order to ensure interoperability, and higher data rates are negotiated, depending on the capabilities of the communicating devices, as part of the process of establishing the link.

Data Link Layer
IrDA's two mandatory protocols, the link access (IrLAP) and link management (IrLMP) protocols, are Data Link (Layer 2) protocols in terms of the OSI model (Figure 10-11).

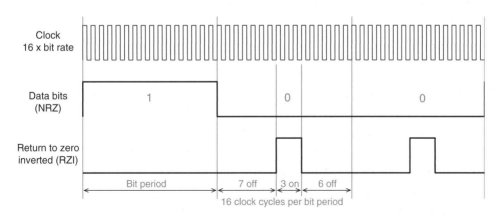

Figure 10-12: IR Pulse Shortening (SIR RZI Modulation)

IrLAP establishes the device to device connection, controlling the discovery and addressing of devices within range and establishing the best common data transmission rate. On discovery, devices randomly choose and exchange 32-bit IrLAP addresses. The devices in an IrLAP connection have a master–slave relationship with the master responsible for sending command frames, initiating connections and transfers, organising and controlling the flow of data, including dealing with data link errors.

The slave device sends response frames, responding to the commands and requests of the master device. Once the IrLAP connection is started, media access is controlled by time division (TDMA) with master and slave taking alternate 500 millisecond time slots.

The distinction between master and slave devices is relevant at the Data Link level. However, at the application level, once a connection is made between two devices, an application on a slave device can initiate an operation on the master device just as easily as the other way round.

IrLMP, the link management protocol, multiplexes services and applications on the connection established by IrLAP. IrLMP also handles address conflict resolution in the case of a new device being discovered which requests the same IrLAP address.

IrDA Optional Protocol Stack
IrDA's optional protocol stack (Figure 10-13) offers applications a number of new services and emulations of legacy services, the most important of which are described below.

LM-IAS
The link management information access service (LM-IAS) provides a database to enable applications to discover devices and to access device-specific information, in essence a "yellow pages" of devices and the services they can provide. All services or applications available via an incoming connection must have an entry in the LM-IAS database, which can be queried to obtain information about these services, for example the current load experienced by a network resource or the attributes of a serial link emulation.

Tiny-TP

Tiny-TP is an intermediate protocol layer that provides a simple transport protocol to control flow on IrLMP connections. It also provides a segmentation and re-assembly service to prevent deadlock situations that can occur as a result of limited device buffer space. Tiny-TP controls flow by adding a single "credit" byte of overhead to each transmitted frame. A receiving device can use this credit when one of its applications need to transmit an LMP frame back to the other device. This simple system is similar to the flow control mechanism used in Bluetooth's RFCOMM protocol, described above, and ensures that communication is not interrupted by a device running out of buffer space.

IrCOMM

IrCOMM emulates legacy serial (or parallel) port connections for applications such as printing. When installed, IrCOMM creates a virtual port which appears to the host computer or applications as if it were a standard serial or parallel port connection. IrCOMM includes emulation of a number of legacy interfaces including RS232 and Centronics LPT.

IrOBEX

IrOBEX is an optional application layer protocol that is designed to enable applications to exchange a wide variety of arbitrary data objects such as files, electronic business cards and digital images. It defines the

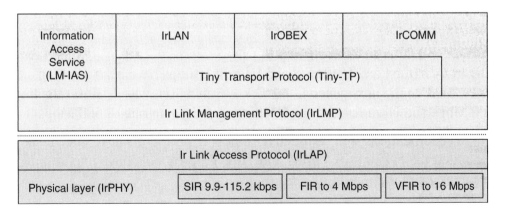

Figure 10-13: IrDA Optional Protocol Stack

conversion of any file into a universal object and also provides tools to enable the object to be understood and handled correctly on the receiving side of the link. IrOBEX serves a similar role to HTTP in the Internet protocol suite.

IrLAN

IrLAN allows devices to access Local Area Networks by emulating a low level Ethernet link, including TCP/IP. Using IrLAN, a computer can attach to a LAN via an access point device (an IrLAN adapter) or through a second computer already attached to the LAN. Two computers can also use IrLAN to communicate as though attached via a LAN, giving each computer access to the other computer's directories and other network resources.

IrDA in Practice

IrDA is the most widespread wireless networking technology in use today. It provides a simple and secure method for transferring files between personal computing and communication devices, and is firmly associated with such applications as PDA to laptop synchronisation, business card and mobile phone data exchange.

Apart from IrDA ports in laptops and PDAs, over 200 million IrDA enabled mobile phones were shipped in 2004. With the increased popularity and pixel count of digital cameras in mobile phones, these IrDA links are also being put to use for direct photo printing and image file transfer.

Current and Future Developments of IrDA

The IrDA IrBurst and UFIR Special Interest Groups have been working since 2003 on the next generation IrDA specification, with IrBurst targeting 100 Mbps and Ultra Fast IR (UFIR) aiming for a data rate of 500 Mbps.

These specifications will also deliver a new Ir protocol stack, since tests show that the existing IrCOMM and Tiny-TP protocols have maximum throughputs in the region of 3 Mbps.

The market driver for these developments is seen as a user demand to transfer compressed video between devices, and the target is to transfer one hour of MPEG2 compressed video (100–200 MB) over a handheld link in no more than 10 seconds. One usage model anticipates that a customer will use a handheld device such as a mobile phone to pay for and download video content from vending machines on the street.

Future developments will also see an extension of the effective range of IrDA beyond the current one metre limit. This will enable the "mobile phone as digital wallet" usage model to be extended out of doors, to applications such as motorway toll collection.

Near Field Communications

Origins and Main Characteristics

Near field communication (NFC) is an ultra short range wireless communication technology that uses magnetic field induction to enable connectivity between devices when they are in physical contact or within a range of a few centimetres. NFC has emerged as a technology for interconnecting consumer electronic devices from the convergence of contactless identification (e.g. RFID) and networking technologies, and aims at simple peer-to-peer networking through automatic connection and configuration.

The key difference between NFC and standard RF wireless communication is the way in which the RF signal is propagated between the transmitter and receiver, as described in the Section "RF Signal Propagation and Reception, p. 105". Standard RF communications, such as a Wi-Fi, is described as "far-field" since the communication range is large compared to the size of the antenna. Near field communication relies on direct magnetic or electrostatic coupling between components within the communicating devices rather than free space propagation of radio waves.

Because of the very short range, NFC devices can communicate using extremely low electric or magnetic field strengths, well below regulatory noise emission thresholds, so that there are no limitations on frequency band usage due to licensing restrictions.

NFC technology is a joint development of Philips and Sony, and is based on the ECMA 340 standard. The technology is being promoted by the NFC Forum, whose sponsor members also include MasterCard, Motorola, Nokia and Visa International.

The ECMA 340 standard was adopted by the ECMA General Assembly in December 2004, and defines NFC communication modes using inductive coupled devices operating at a centre frequency of 13.56 MHz. The definition is also known as the near field communication interface and protocol (NFCIP-1). Similar to the more familiar IEEE standards, ECMA 340 specifies the modulation and data coding schemes, data rates and frame format for NFC device interfaces. A simple link layer protocol addresses link initialisation and collision avoidance, and a transport protocol, covers protocol activation, data exchange and deactivation.

NFC PHY Layer

ECMA 340 specifies a magnetic induction interface operating at 13.56 MHz and with data rates of 106, 212 and 424 kbps, which is compatible with Philips' MIFARE® and Sony's FeliCa™ contactless smart card interfaces.

Rather than measuring transmitter power and receiver detection levels in dBm as is the case for far-field RF communication, the strength (H) of the magnetic field used in NFC is measured in amps/metre (A/m). ECMA 340 specifies the field values as shown in Table 10-17.

The ECMA 340 standard defines two communication modes — active and passive. In the active mode, communication is started by an RF field generated by the initiating device (the Initiator) and the target device (the Target) also generates a modulated RF field to respond to the Initiator's command. Modulation and bit coding methods used in active mode are shown in Table 10-18.

In the passive mode (Table 10-19), the Initiator starts the communication using an RF field but the Target responds by load modulation rather than by generating an RF field in response. Load modulation, described in the

Table 10-17: ECMA 340 NFC Magnetic Field Strength Specification

Field level	Field strength	Description
$H_{threshold}$	0.1875 A/m	Minimum field detection level
H_{min}	1.5 A/m rms	Minimum un-modulated field strength
H_{max}	7.5 A/m rms	Maximum un-modulated field strength

Table 10-18: ECMA 340 Active Mode Modulation and Bit Coding Methods

Bit rate	Modulation method	Bit coding method
106 kbps	ASK (100% modulation)	Pulse position coding (Modified Miller) — pulse transmitted at the centre of a bit period for each 1-bit, or at the start of a bit period for an opening 0-bit or a repeated 0-bit.
212/424 kbps	ASK (8–30% modulation)	Manchester coding — transition at the centre of each bit period; low to high for a 0-bit, high to low for a 1-bit. Reverse polarity (i.e. high to low for a 0-bit, low to high for a 1-bit) is also allowed.

Section "Load Modulation, p. 127", entails modulating the load in the target device that the initiating RF field is applied to. This generates sidebands on the original carrier frequency (13.56 MHz) that are detected by the Initiator.

Protocol Stack

Since NFC is not attempting to provide the full range of network features captured in the OSI model, the protocol stack is very limited and consists of a single simple transport protocol, which defines activation, data exchange and deactivation on an NFC link.

The vestiges of a Data Link layer are also evident in the form of media access control based on CSMA/CA. An Initiator checks for an existing

Table 10-19: ECMA 340 Passive Mode Modulation and Bit Coding Methods

Bit rate	Modulation method	Subcarrier frequency	Bit coding method
106 kbps	Load modulation	$f_c/16 = 847.5$ kHz	Subcarrier modulated using Manchester coding. Reverse polarity not allowed.
212/424 kbps	Load modulation	-	Carrier modulated using Manchester coding. Reverse polarity allowed.

RF field before commencing communication and similarly a Target device in active mode checks for an existing RF field before responding.

A single initiating device can interact with multiple targets, each of which generates a random 40 bit ID at the start of the device selection process. The discovery of target device IDs involves an elegant process to resolve collisions which will occur when several targets respond at the same time, particularly when targets are responding in passive mode (Figure 10-14).

Collision detection at the bit level is made possible by the use of Manchester coding, since a collision is detected when a full bit period occurs without a transition being sensed. This can only occur when a 1-bit transmitted by one target collides with a 0-bit transmitted by another target. Bits received before the collision can be recovered and the targets are requested to re-send data starting with the unrecovered bit. A random delay used by responding targets ensures that this process does not get stuck in a repeating loop.

The data link between devices is transaction based, with initiation and termination occurring around a single data transfer. Initiator and Target negotiate a communication speed, starting with the lowest (106 kbps), in a parameter selection step during transport protocol initiation.

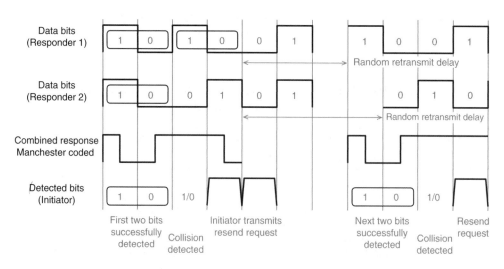

Figure 10-14: NFC Collision Detection with Multiple Responding Targets

NFC in Practice

Four basic NFC usage models are currently envisaged, as shown in Table 10-20.

Apart from these usage models in which the NFC connection is used to transfer end-user data, NFC can also be used to securely initiate another connection between two NFC enabled devices. For example, NFC

Table 10-20: NFC Usage Models

Usage model	Description	Example
Touch and go	The user brings the device storing a ticket or access code close to the reader for applications such as event or transport ticketing and access control, or for simple data capture, such as picking up an Internet URL for further information from a smart label on a poster or other advertising.	You see a poster advertising an event such as a concert you want to attend. Bring your PDA or mobile phone near the poster to download event information from a smart chip in the poster.
Touch and confirm	Transactions such as mobile payment where the user is required to enter a password or other confirmation to authorise the interaction.	Event tickets could be purchased online or from an electronic box office and stored on your handheld device.
Touch and connect	Two NFC-enabled devices can be linked to enable peer-to-peer data transfers, such as exchanging photos or synchronising contact information.	If you take pictures with your mobile phone's built in camera, you can later touch an NFC enabled computer or TV to display the images, or touch and transfer them to a friend's mobile phone.
Touch and explore	NFC enabled devices may offer a range of possible functions, including other high speed connectivity options. The simple NFC connection will allow the user to explore a device's capabilities and access other available services or functionality.	By simply touching two devices together it will be possible to transfer large files between the devices, for example, using NFC to identify and configure a separate high-speed wireless connection.

enabled Bluetooth or Wi-Fi devices may use NFC to initiate and configure the longer range link. Security is assured by the close proximity requirement for NFC operation. Once the Bluetooth or Wi-Fi link is configured, the devices can be separated for longer range communication.

Current and Future Developments of NFC

The first examples of NFC in use have been trials followed by commercial deployment for transport ticketing and payment on a local bus network in Hanau, Germany and on Taipei's Mass Transit Rail system in Taiwan. These trials have been based on the Nokia NFC shell, which clips on to a Nokia 3220 mobile phone.

Future data rates up to 1.7 Mbps are currently planned, approaching the 3 Mbps of Bluetooth 2.0, and market research points to 50% of mobile phones being NFC enabled by 2010.

Summary

The simple PAN landscape, dominated by IrDA and Bluetooth, is becoming increasingly diverse, as shown in Figure 10-15, with many new technologies being developed that will offer the user a wide range of choices in terms of data throughput, range, power consumption and battery life.

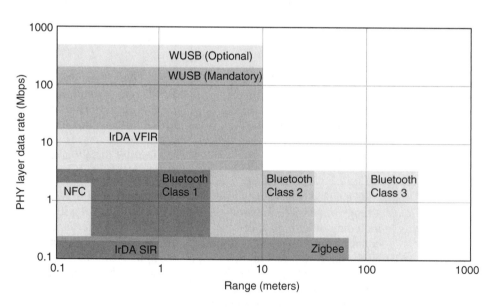

Figure 10-15: PAN Technologies; Range vs. Data Rate

Several of these technologies, such as ZigBee and NFC, have specific niche applications and in the short term the main choice for general personal area networking will be between Bluetooth and one of emerging high speed UWB radio based technologies such as wireless USB.

Looking to the future, the IEEE 802.15 working group has chartered a standing committee (IEEE P802.15 SCwng) to look at the technologies that will lead to the next generation of wireless PANs.

The next chapter looks at some of the considerations that impact on the choice of PAN technology for a given application, and some further practical aspects such as the security and vulnerabilities of the various PAN technologies.

Implementing Wireless PANs

Wireless PAN Technology Choices

The task of planning and implementing a wireless PAN is significantly simpler than the process discussed in Chapter 7 for wireless LANs, but the same three initial steps can also be applied here:

- Establish the user requirements; what is it that the user wants to be able to achieve with the PAN and what are the expectations of performance?

- Establish the technical requirements; what attributes does the technological solution need to possess in order to deliver these user requirements?

- Evaluate the available technologies; how do each of the available or emerging PAN technologies rank against the technical requirements?

Establishing User Requirements

User requirements are independent of specific technologies and should be expressed in terms of the user experience rather than any particular solution or technical attribute. For example, in relation to battery life for mobile devices, power consumption is a technical attribute, whereas the length of time between battery recharging is what the user is really concerned about.

For a PAN implementation project targeting a large user group it will be important to gather a wide range of views on user requirements — for example using a questionnaire or by interview. As a first step it may be necessary to raise awareness by demonstrating the technology to the prospective user group, so that they are better able to give an informed view on requirements.

Common types of user requirement are listed and discussed in Table 11-1.

Table 11-1: PAN User Requirement Types

Requirement type	*Considerations*
Usage model	It is important to be clear what types of use the PAN will be put to; portable device synchronisation with a desktop or laptop computer, data exchange between portable devices, etc. Is the usage model likely to change in future or are requirements well defined and static?
Device types	What type of devices will be used in the PAN? Examples include laptop computer, PDA, mobile phone, hands free headset, personal video player. There may also be a requirement to connect to non-mobile devices — a desktop computer or LAN and related resources.
Performance expectations	What are the user's performance expectations? This will be particularly important if the usage model includes the transfer of large data files or media streaming.
Device weight and size	Particularly for PAN devices that are worn, such as a hands-free headset, minimising device size and weight is likely to be a requirement.
Ease of connection	How important is ease of connection and reconnection. If a device will be used often but intermittently, then the user may not want to perform port activation and authorisation each time a connection is made, such as is required for an IrDA link. Bluetooth's preferred device mode would meet this requirement.
Mobility	Is the network required to be moveable between locations (portability) or to operate while the user is physically moving? For example, IrDA is not suitable if any device is mobile in view of the need for port alignment.

Table 11-1: PAN User Requirement Types — cont'd

Requirement type	Considerations
Device interoperability	The required functionality must be implemented in both devices to enable interoperability. For example, a mobile phone may have a headset profile implemented to inter-operate with a Bluetooth headset, but will not provide Internet connectivity if the dial-up networking profile is not implemented.
Operating environment	Are there specific requirements in the environment where the PAN will operate, for example will it need to operate in the presence of narrow band or other RF interference, perhaps from a co-located WLAN?
Battery life	How often will the user need to recharge battery operated devices? Are features to conserve battery power easy to access and configure?
Cost	It goes without saying that the user will want value for money in the chosen solution, but other soft issues may play a part here too. Particularly for PAN equipment, the aesthetic aspects of personal accessories may also be an implicit or explicit requirement.

Establishing Technical Requirements

Technical requirements follow from user requirements, translating these into the related technical attributes (Table 11-2). For example, if a user requirement is Internet access from a mobile device, then the required technical attribute is an IP networking capability. Alternatively, a user

Table 11-2: PAN Technical Attributes

Requirement type	Considerations
Application support	Does the technology support the specific usage models required by the user?
Effective data rate	The required data rate will be dictated by the usage model, specifically the typical size of data objects that will be transmitted across the PAN, and the user's performance expectations in terms of upload/download time.

Continued

Table 11-2: PAN Technical Attributes — cont'd

Requirement type	*Considerations*
Quality of service	If the usage model includes applications that require isochronous data transmission, then guaranteed quality of service will be an important attribute to ensure performance expectations are met.
Interference and coexistence	If the PAN will have to operate in an environment with other wireless networks (e.g. an IEEE 802.11 WLAN) then coexistence will need to be a consideration.
Power consumption	Using PAN features on a mobile device may significantly increase power consumption, and detract from the overall performance of the device by reducing battery life. Network features, such as actively searching for other devices, should be easy to deactivate when not required in order to maximise battery life, for example on a Bluetooth enabled PDA.
Operating system and other software compatibility	Operating system compatibility will be an issue when applications attempt to compatibly inter-operate over the PAN link. Additional software components may be required, for example to exchange data between a mobile phone and PDA.
Technology maturity	Considerations will vary with the stage of maturity; before standards have been agreed, early products have an interoperability risk; a fully mature technology may have limited scope for future development and risk early obsolescence as new usage models arise.
Operating range	This is less important for a PAN, since by definition only a limited operating range is required — the personal operating space. However, PAN technologies vary widely in their achievable range — from 0.1m for NFC to over 100m for ZigBee.
Cost	With an increasing range of capabilities coming to market in the next few years, significant price differentiation between PAN technologies can be expected. If options like ZigBee and NFC meet the user requirements they will be considerably cheaper than Bluetooth or the UWB alternatives.

requirement to stream video to a handheld media player will translate into technical attributes for QoS and very high data rate.

Evaluating Available Technologies

Having established the technical attributes necessary to meet user requirements, the available technologies can then be directly assessed against these attributes. A simple table can be used to display the comparison, similar to the example shown in Table 11-3.

Although the range of available PAN technologies is growing, the choice is still sufficiently narrow not to require sophisticated evaluation methods, such as assigning a relative weight to the various requirements.

This approach will result in a transparent and objective comparison of the available solutions, but an independent reality check will always be helpful to verify the proposed solution — to ensure that no requirements have been missed and no limitations of a particular technology have been overlooked.

It is also helpful to research the solutions that have been adopted by others to meet similar needs. If no examples can be found of the proposed solution being used in practice (IrDA for last mile broadband access?) then either a new technology application has been spotted or something has been missed in the evaluation.

Table 11-3: PAN Technologies; Technical Attribute Comparison

Requirement type	*Bluetooth*	*WUSB*	*ZigBee*	*NFC*	*IrDA*
Effective data rate	ca. 3 Mbps (2.0)	480 Mbps	250 kbps	ca. 2 Mbps	16 Mbps (VFIR)
Quality of service	–	++			–
Interference and coexistence	–	++	–	++	++
Power consumption	High	Low	Very low	Very low	Low
Technology maturity	Very mature	New	New	New	Very mature
Operating range	< 10 m	10–30 m	70–300 m	< 0.2 m	1 m

Pilot Testing

If the PAN implementation project is targeting a significant number of users then, as for WLAN implementation, a pilot testing phase will be beneficial. This will be the opportunity to confirm that the statement of user requirements is clear and complete, and that the users' performance expectations are also well defined and can be met by the proposed solution.

The group of users chosen for the pilot test should cover the full range of technical capabilities present in the final user group — from the most technically savvy, whose performance expectations will be hardest to meet, to those who may be less demanding on performance but will have a natural focus on ease of use. Taking account of a wide diversity of views on how well the solution meets the full range of user requirements may make the implementation task more challenging, but will ultimately result in wider user acceptance of the end result.

Wireless PAN Security

Although a wireless PAN will generally have a more limited range than a WLAN (with the possible exception of ZigBee networks), ensuring security will remain an important implementation issue, since most wireless PAN technologies are potentially vulnerable to a variety of security threats.

The following sections summarise the security features and known vulnerabilities of various PAN technologies, and provide guidance on security set-up during implementation.

Bluetooth Security

Bluetooth Security Overview
Bluetooth includes comprehensive security measures designed to ensure that access to services is protected and only granted to another device after appropriate authorisation. Three types of service security levels are defined, as shown in Table 11-4.

The first step in the process of establishing a secure Bluetooth connection is authentication, which occurs after the initial pairing and results in the

Table 11-4: Bluetooth Service Security Levels

Security mode	*Service type*	*Security level*
1	Open services	These services can be accessed by any device. There are no security requirements and authentication and encryption are bypassed.
2	Authentication only services	These services can only be accessed by authenticated devices.
3	Authentication and authorisation services	These services may only be accessed by trusted devices.

creation of a semi-permanent authentication key that is shared by the two devices.

Authorisation is the second step, and may be required before a device will give another device access to a requested service. Authorisation can be completed without user intervention if the requesting device is marked as "trusted". Trust is normally granted to a device by the user during an initial authorisation.

Three levels of device security are also defined, as shown in Table 11-5.

As a third security step, transmitted data can be encrypted using a key generated from the existing authentication key. The maximum length of the encryption key, up to 128-bit, is negotiated between master and slave

Table 11-5: Bluetooth Device Security Levels

Device type	*Security level*
Trusted devices	Devices which are identified in the security database as trusted and are allowed unrestricted access to all services.
Known untrusted devices	Devices which have been paired and perhaps authenticated but which are not identified in the database as trusted. Access to certain services may be restricted.
Unknown devices	Devices which have not been paired and for which no security information is known. Only open services are accessible to unknown devices.

as part of the process of initiating encryption. Although it is impossible to prevent the interception of data transmitted by radio, the use of FHSS makes Bluetooth virtually immune to eavesdropping by a device that does not follow the same hopping pattern.

Bluetooth Vulnerabilities

Provided that security modes above 1 are enabled and reasonably long passphrases or PINs are used, Bluetooth security generally prevents unauthorised access to data or services on enabled devices. However, there are two known vulnerabilities, "bluesnarfing" and "bluebugging" that affect some mobile phones, although software upgrades have been developed by vendors for phones affected by these vulnerabilities. "Bluejacking" is not strictly a security vulnerability, but represents a subversion of the normal pairing process that may lead to undesired pairing to another device.

Bluesnarfing enables a hacker to access data stored on a Bluetooth mobile phone without alerting the phone's user that a connection is being made. Phonebook, calendar, and any other data stored in the phone's memory can be accessed in this way.

Bluebugging allows a hacker to access the mobile phone commands without alerting the phone's user. This allows the hacker to make phone calls, send text messages, read and write address book information and listen in on phone calls.

Bluejacking uses the first step of the pairing process to send a message to another Bluetooth mobile phone by entering the message in the 248-character field that would usually contain the name of the initiating device. The message may invite the recipient to enter a response that has been selected by the sender as the passphrase, resulting in an undesired pairing.

Both bluesnarfing and bluebugging can be made more difficult, although not impossible, by setting the mobile phone in non-discoverable mode, and bluejacking is not possible if the phone is in this mode.

Bluetooth Security Measures

There are a number of measures that can be taking when operating Bluetooth devices to ensure the security of data stored on devices or transmitted between them, as shown in Table 11-6.

Table 11–6: Bluetooth Security Measures

Security measure	Description
Secure pairing location	Bluetooth security is at its most vulnerable at the initial pairing stage when the passphrase or PIN is being entered into the pairing devices. A request to enter a PIN code should only be responded to when it is part of a desired pairing being conducted in a secure environment.
Non-discoverable state	Bluetooth devices will be less vulnerable if they cannot be discovered by potential attackers. Set devices to a non-discoverable state for routine use. Make devices discoverable, in a secure environment, when necessary to establish new connections.
Maximum PIN length	Using the minimum 4 character PIN makes it easier for a hacker to intercept. This vulnerability can be eliminated by using PINs of at least 8 characters and preferably of the maximum allowed length. Make sure no devices are using a default PIN.
Security mode	Use authentication and encryption (Security Mode 3) for any confidential communications. In a multiple hop link, ensure that all links in the communication chain are using the required security mode.
Anti-virus software and security updates	Anti-virus software can be installed on many Bluetooth devices in the same way as on a personal computer. Anti-virus software, as well as device operating software, should be kept up to date with manufacturers' revisions and security updates.
Software downloads	Only download or install software from trusted sources. Careful attention should be given to any security warning during software installation.
Unpair from lost devices	If a Bluetooth device is lost or stolen, it should be unpaired with all devices it was previously paired with, by deleting the lost device from the list of paired devices on these devices. Failure to do this will make these devices vulnerable to attack by the previously paired device.

Wireless USB Security

The goal of wireless USB security is to provide the same level of confidence that the user has when making a wired USB connection, namely that the devices connected are only those that the user wants to be connected and that the transmitted data is protected from unwanted external observation or modification.

Table 11-7: Fixed Symmetric Key Authentication Steps

Authentication step	*Description*
Device distributes its FSK to the user	This key may be either printed on the device or included in the installation software.
User transfers the FSK to the host	User confirms trust of the device carrying this FSK and instructs the host to allow this new connection.
Host confirms to device that a new connection is allowed	By its knowledge of the device key the host is able to demonstrate to the device that it too has the user's trust.
User instructs device to start new connection	This may be for example by pushing a "Connect" button on the device.

Authentication will include the manual entry or confirmation of a connection key (a PIN or passphrase) by the user, to ensure that hosts and devices are mutually able to demonstrate user trust when requesting or allowing a connection. Three different types of authentication "ceremony" are possible, depending on whether the connection key is distributed directly by the user to host and device, is hard wired into the device, or is based on an exchange of public keys between host and device.

The authentication ceremony that most closely mirrors the process of making a wired USB connection is the second of these, based on sharing a Fixed Symmetric Key (FSK) that is typically hard wired into the device at manufacture. The steps in this authentication ceremony are shown in Table 11-7.

The ceremony for public key authentication will be similar, except that both device and host will present their public keys to the user as part of a software driven installation process.

After these authentication steps, association continues with a handshake process resulting in the generation of a 128-bit AES encryption key that is used to protect the connection. This pair-wise temporal key (PTK) is unique to the connection and is used by host and device to encrypt all transmitted and decrypt all received data packets.

ZigBee Security
The IEEE 802.15.4 MAC layer specifies four services that are available to implementers in the ZigBee security software toolbox to ensure security

Table 11-8: IEEE 802.15.4 MAC Security Services

Security service	Description
Access control	The network co-ordinator acts as a "trust centre", maintaining overall network knowledge, including a list of trusted devices within the network, as well as maintaining and distributing network keys.
Data encryption	Optional 128-bit AES using link keys between devices or a common network key.
Frame integrity check	Check to ensure that data transmitted within a frame has not been modified
Sequential freshness check	A sequentially updated freshness value that allows the network controller to check for and reject any replayed data frames.

and data integrity, as shown in Table 11-8. Specific security implementations can then be developed using these services.

Two security modes are defined in the ZigBee standard, commercial and residential. The full access control functionality of the network co-ordinator, or trust centre is only available in the commercial security mode. In residential mode the network co-ordinator controls device access to the network but does not establish or maintain keys, in order to reduce the memory cost of the trust centre device.

IrDA Security

The IrDA standard does not include a specification of link level security as the short range and line-of-sight requirements provide an inherent low level security. Any threat of unauthorised access to data through an active port, for example when using an IrDA enabled laptop in a public setting, can be easily countered by disabling the IrDA port when not in use and by ensuring that sensitive data is only transferred in a private environment.

Additional security measures, such as authentication and encryption, are implemented at the application level. One such example is the IrDA OBEX authentication mechanism, which requires a user-entered OBEX password to be stored on both devices before an OBEX connection can be made. This password is then used by both devices for authentication when the link is established.

Summary of Part IV

In common with wireless LANs, wireless PAN implementation can also benefit from a systematic approach to establishing requirements and selecting the most appropriate technology. Pilot testing can also be a valuable implementation stage, particularly if the WPAN is being deployed for a large and diverse user group.

Despite their inherently shorter range, WPANs are also susceptible to a range of security threats and appropriate security measures should be considered, in line with the level of risk to user data from unauthorised access.

PART V

WIRELESS MAN IMPLEMENTATION

Introduction

Although Wi-Fi has provided a basis for many small to medium scale metropolitan area networking initiatives, it is only with the completion of standards specifically aimed at providing wireless "last mile" solutions, such as IEEE 802.16, that the prospect has opened up of more widespread MAN applications.

The key requirements for metropolitan area networking are;

- Scalability to hundreds or thousands of subscribers rather than perhaps tens or hundreds of users on a LAN

- Flexibility to provide access for a wide range of different service types, including mechanisms for requesting and allocating bandwidth

- Guaranteed quality of service (QoS) when required by individual subscribers or services.

Despite the recent advances discussed in Chapter 6, LAN standards such as IEEE 802.11 still fall short of providing these requirements, hence the need for specific standards to address the requirements for metropolitan area networking.

In Chapter 12 the IEEE 802.16 standards will be described, focussing on how these key requirements have been achieved. 802.16's European sibling HIPERMAN, which has been developed by ETSI alongside the IEEE standard, will also be briefly covered.

Chapter 13 will address MAN implementation, covering the design and start-up of a wireless metropolitan area network or its rural equivalent. This will cover technical planning and implementation aspects — site surveying, equipment planning and installation — and also the business planning aspects — customer mapping, competitor analysis and management and financial planning.

CHAPTER **12**

Wireless MAN Standards

The 802.16 Wireless MAN Standards

Origins and Main Characteristics

The IEEE 802.16 set of standards have been developed since 1998 in response to the need for a wireless solution to supplement xDSL and cable modems in delivering broadband access to homes and small businesses. A wireless solution for "last mile" broadband access has the advantage of being able to provide wide geographical coverage with minimal infrastructure cost, and therefore brings with it the potential to accelerate broadband uptake.

The evolving suite of IEEE 802.16 standards is shown in Table 12-1. The 10–66 GHz frequency range was the initial focus of the IEEE Working Group on broadband wireless access (BWA) which developed the standards. This was largely motivated by worldwide spectrum availability in this frequency range, and led to the approval of the initial IEEE 802.16 standard in 2001 and its publication in March 2002.

The Working Group then turned its attention to the 2–11 GHz range, where the advantages of lower cost implementation and non line-of-sight transmission outweighed the potential difficulties in this crowded piece of RF spectrum. The result was the IEEE 802.16a standard, approved in 2002 and published in January 2003.

IEEE 802.16 has been designed to offer considerable flexibility in the specification of the PHY layer, in order to accommodate varying requirements (such as channel widths) in different regulatory regimes.

Table 12-1: The IEEE 802.16 Standard Suite

Standard	Key features
802.16	Original standard, approved in 2001. 10–66 GHz spectrum, line-of-sight links up to 134 Mbps.
802.16a	Approved in 2002. 2–11 GHz. Non line-of-sight links up to 70 Mbps.
802.16b	Update to 802.16a dealing with unlicensed applications in the 5 GHz band.
802.16c	Update to 802.16 addressing interoperability of 10–66 GHz systems.
802.16d	The basis of "WiMAX", replaces 802.16a and includes support for advanced antenna systems (MIMO). Approved in June 2004 as 802.16-2004.
802.16e	Extension to provide mobility, including rapid adaptation to the changing propagation environment.
802.16f	Extension to support multi-hop capabilities required for mesh networking.
802.16g	Addresses efficient handover and QoS for mobile networking.
802.16h	MAC enhancements to enable coexistence of licence exempt 802.16 based systems and primary users in licensed bands.

These different air interfaces are supported by a common MAC layer, which has been designed to provide the key requirements for metropolitan area networking — scalability, service type flexibility and quality of service.

802.16 PHY Layer

PHY Layer for 10–66 GHz Spectrum

In this extremely high frequency range, RF propagation requires for all practical purposes a line-of-sight between transmitter and receiver. Given this limitation, it is not necessary to consider the use of complex techniques such as OFDM to overcome multi-path effects, which occur in a non-line-of-site environment, and the Working Group therefore selected a simple single carrier (SC) modulation technique for this interface (Table 12-2).

In downlink transmissions (from the base station (BS) to subscriber stations (SS)), time division multiplexing (TDM) is used, with time slots

Table 12-2: 802.16 Key Parameters

Parameter	*802.16 standard*
RF band	10–66 GHz
Modulation	Single carrier modulation (SC) (QPSK, 16- & 64-QAM)
Data rate	Peak data rates to 134 Mbps
Channelisation	20, 25 or 28 MHz channel widths
Duplex method	TDD and FDD, as well as half-duplex via TDMA
Network topology	Point-to-multipoint
Bandwidth allocation	Grant per subscriber station (GPSS) — see the Section "802.11 PHY Layer, p.310 "

allocated to individual subscribers. This enables bandwidth to be guaranteed to latency sensitive services. In the uplink direction (from SS to BS) time division multiple access (TDMA) is used.

Duplexing of up and downlinks can be achieved either by time division or frequency division duplexing (TDD or FDD). Half-duplex subscribers, transmitting and receiving on the same channel, are also supported by

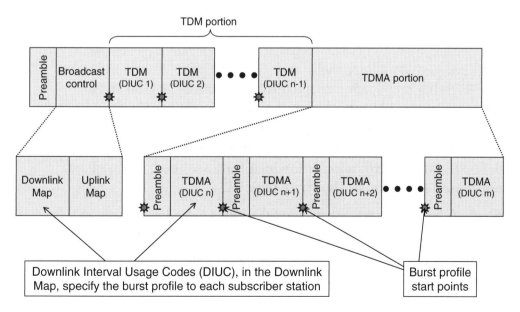

Figure 12-1: 802.16 Downlink Frame Structure

having an optional TDMA uplink segment following the TDM portion of each downlink data frame.

A range of modulation and coding schemes are available (including QPSK, 16-QAM, 64-QAM) and each subscriber station negotiates a scheme with the base station, in line with its particular needs for efficiency (depending on data rate) and robustness (depending on signal propagation environment/signal strength).

The result is that, within a single downlink frame transmitted by the base station (Figure 12-1), individual data bursts destined to individual subscriber stations will be transmitted with different coding and modulation schemes — a technique known as adaptive burst profiling. Achievable data rates for different modulation methods are shown in Table 12-3.

Table 12-3: 802.16 Bit Rate vs. Channel Width and Modulation Method

Channel width (MHz)	*Bit rate (Mbps) per modulation method*		
	QPSK	16-QAM	64-QAM
20	32	64	96
25	40	80	120
28	44.8	89.6	134.4

PHY Layer for 2–11 GHz Spectrum

The differing propagation characteristics in the 2–11 GHz range, compared to the EHF range up to 66 GHz, require an air interface that can accommodate significant multi-path propagation in a non line-of-site operating environment. Three optional PHY layer specifications are defined in the 802.16a standard, which addresses both licensed and unlicensed spectrum, as shown in Table 12-4.

Most of the key parameters of the 802.16a standard are common to its higher frequency predecessor, as shown in the Table 12-5.

In addition, 802.16a also supports advanced antenna systems. As described in the Section "Wireless LAN Antennas, p. 55", advanced antenna systems or smart antennas can improve link robustness by suppressing interference and increasing overall system gain. As the cost

Table 12-4: 802.16a Optional Air Interfaces

Air interface	Summary
WirelessMAN SC2	A single carrier modulation format, providing interoperability with the 10–66 GHz single carrier air interface.
WirelessMAN OFDM	Orthogonal frequency division multiplexing, with TDMA controlled multiple user access. This interface is mandatory for unlicensed bands.
WirelessMAN OFDMA	An OFDM interface with user access controlled by allocating to individual users a subset of the full set of available carrier frequencies.

of these systems comes down, they will play an important role in improving wireless network performance, particularly in the increasingly crowded unlicensed frequency bands.

MAC Layer
In fulfilling the needs of metropolitan area networking, the 802.16 MAC layer must provide flexible and efficient access for a wide range of different service types. The main shortcoming of the contention based

Table 12-5: 802.16a Key Parameters

Parameter	802.16a standard
RF band	2–11 GHz
Modulation	Single carrier, OFDM
Data rate	Peak data rates to 70 Mbps
Multiple access	OFDMA, TDMA
Channelisation	Flexible channel widths, from 1.75 to 20 MHz
Duplex method	TDD and FDD, as well as half-duplex via TDMA
Network topology	Point-to-multipoint and mesh topologies
Bandwidth allocation	Grant per subscriber station (GPSS) or grant per connection (GPC) — see Section 12.1.3 "MAC Layer"

media access of 802.11 networks, based on carrier sensing (CSMA-CA), is that, prior to the 802.11e enhancements, no particular service quality level could be guaranteed. A subscriber requiring a latency-sensitive service such as VoIP may be subject to the exposed station or hidden station problems (Section "Radio Transmission as a Network Medium, p. 75") with the consequent deterioration of service quality.

Connection-Oriented Versus Connectionless
One key to the effectiveness of the 802.16 MAC is its connection-orientation. Every service is mapped to a connection and referenced using a 16-bit connection ID (CID). This includes services which are inherently connectionless such as UDP services (for example RIP, SNMP or DHCP messaging). Each connection can then be associated with specific parameters such as;

- bandwidth granting mechanism (continuous or on-demand)

- Associated QoS parameters

- Routing and transport data.

Connections are typically unidirectional, so that different QoS and other transport parameters can be defined for the uplink and downlink directions.

When a new subscriber station joins an 802.16 network, three connections are initially opened to carry management level messages, as shown in Table 12-6.

Table 12-6: 802.16 SS Management Connections

Connection	*Usage*
Basic management connection	Short, time-critical MAC and radio link control (RLC) messages.
Primary management connection	Delay tolerant messages e.g. authentication, connection set-up.
Secondary management connection	Standards based management messaging; DHCP, SNMP, RIP.

Additional connections are allocated to subscribers when specific services are contracted, typically in uplink plus downlink pairs. The MAC also reserves a number of connections for other general purposes, such as initial contention-based access and broadcast or multicast transmissions, including polling of SS bandwidth needs.

Radio Link Control

Radio link control (RLC) is another key element of the 802.16 MAC that is required to provide adaptive burst control as well as the more traditional power adjustment functions (TPC).

When a subscriber joins the network, SS and BS exchange messages using the basic management connection to establish initial settings for transmit power and timing. The SS will also request a specific initial burst profile, defining signal modulation parameters, based on equipment capabilities and downlink signal quality. The RLC will continue to monitor signal quality after this initial set-up, and either the SS or BS may request a more robust burst profile if environmental conditions deteriorate (e.g. temporarily switching from 64-QAM to 16-QAM) or a more efficient profile if lower robustness can be tolerated as conditions improve.

The uplink burst profile is under the direct control of the BS, and this control is achieved each time the BS allocates bandwidth to a SS by also specifying the burst profile to be used by the SS. The downlink profile is also controlled by the BS, although changes are at the request of the SS, which alone can monitor the strength of the received signal.

Bandwidth Allocation

Flexible allocation of available bandwidth among subscribers and connections is a third key element of the 802.16 MAC. The variety of services and scalability to potentially hundreds of subscribers per BS will clearly put heavy demands on the efficient use of bandwidth.

The bandwidth requirements of individual subscribers are established when a connection is made and a number of messaging options in the standard allow SSs to request additional uplink bandwidth, to inform the BS of total bandwidth needs or allow the BS to poll individual SSs or multicast groups to establish these requirements. These mechanisms ensure efficient use of available bandwidth and flexibility to

Table 12-7: 802.16 GPC and GPSS Classes of Subscriber Station

SS class	*Capabilities*
GPC subscriber station	Bandwidth granted to the SS can only be used for the connection that requested it.
GPSS subscriber station	Bandwidth granted by the BS need not be used only for the connection requesting it. The SS can use granted bandwidth for any of its connections.

accommodate a wide variety of services or to respond to changing requirements of individual services.

The MAC defines two classes of SSs (Table 12-7) — Grant per connection (GPC) and grant per subscriber station (GPSS), which differ in terms of the flexibility available to the SS in using its allocated bandwidth.

At the cost of some additional complexity, GPSS provides greater efficiency and scalability than GPC, for example allowing a SS to respond more quickly to changing environmental conditions. GPSS requires additional intelligence in the SS in order to manage the QoS of its connections, but is clearly one aspect of the autonomy that has to be delegated to subscriber stations in a mesh network.

Mobile WiMAX

The 802.16 Task Group TGe addressed the challenge of adding mobility to the 802.16 standard. A study group on Mobile Broadband Wireless Access was set up in 2002 with the aim of providing mobile access at vehicular speeds of up to 125 km/hr using licensed frequency bands. The resulting 802.16e standard, also designated 802.16-2005, was ratified by the IEEE in December 2005 and specifies new modulation and multiple access schemes to enable mobile non line-of-sight operation. 802.16e enhances the 802.16's original OFDMA air interface by adding a number of new capabilities, as summarised in Table 12-8.

The two key concepts underlying the 802.16e enhancements are the fixing of subcarrier spacing independently from channel width (scalable OFDMA) and the use of subchannelization to enable range versus capacity trade-offs.

Table 12-8: 802.16e Enhancements

Enhancement	*Capabilities*
Constant subcarrier spacing	Increased resistance to multipath fading and Doppler spread in mobile transmission is achieved by keeping subcarrier spacing constant, independent of channel width. This results in a "scaling" of the number of subcarriers with channel width.
Improved indoor penetration	A subset of the available OFDM carriers is used at higher individual power (subchannelisation) to improve indoor reception.
Flexibility in coverage versus capacity	Subchannelisation in the downlink allows a flexible trade-off between data capacity and operating range.
Advanced antenna support	NLOS coverage and performance are improved using advanced multi-antenna diversity and adaptive antenna systems as well as MIMO radio technology.
Enhanced error correction	New coding techniques are used to improve mobile and NLOS performance (e.g. turbo coding and low-density parity check (LDPC)).
Faster error recovery	Using hybrid-automatic retransmission request (hARQ) to improve error recovery.

Scalable OFDMA

As described earlier, 802.16a specifies an OFDM air interface with flexible channel widths from 1.75 to 20 MHz. Each channel is divided into 256 subcarriers (OFDMA256), so that the subcarrier spacing depends on the channel width and varies from 6.8 to 78.1 kHz.

In mobile applications, varying Doppler shift and multipath delays lead to a degradation of performance, in terms of *SNR* and *BER*, particularly for subcarrier spacings at the low end of this range. Conversely, channel capacity can be increased by using more subcarriers at higher channel widths.

Scalable OFDMA addresses these issues by using a constant subcarrier spacing of 11.2 kHz and varying the number of subcarriers depending on the channel width, from 128 for a 1.25 MHz channel to 2048 for a 20 MHz channel. This maximises channel capacity for the wider channels and ensures that all channel widths are equally tolerant of the delay spread resulting from moving stations.

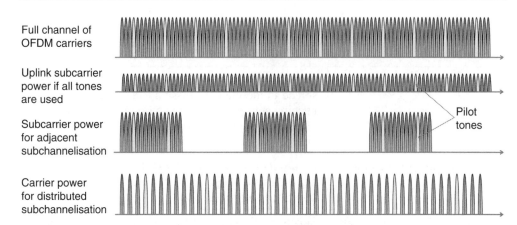

Full channel of
OFDM carriers

Uplink subcarrier
power if all tones
are used

Subcarrier power
for adjacent
subchannelisation

Pilot
tones

Carrier power
for distributed
subchannelisation

Figure 12-2: WiMax Subchannelization

S-OFDMA and OFDMA256 will not be compatible, so that WMAN equipment will have to be replaced in order to support mobile applications based on S-OFDMA.

Subchannelization

Subchannelization refers to the use of a subset of available OFDM subcarriers. By concentrating the available transmitter power into fewer subcarriers, each subcarrier can be transmitted with higher individual power. This additional link margin can be used either to extend the link range, to overcome penetration losses allowing mobile devices to be located indoors, or to reduce the power consumed by the transmitting device.

These benefits come at a cost of reduced link capacity, since only a subset of subcarriers are used to carry data, allowing a trade-off between throughput and mobility.

Various methods can be used for allocating subcarriers to subchannels — the main ones being the use of adjacent or distributed subcarriers (Figure 12-2). Distributed subcarrier allocation is used for mobile applications mainly because the use of a wide range of frequencies (frequency diversity) makes the link less susceptible to the rapidly varying fading that is characteristic of mobile applications.

Subchannelization is also an option in fixed WiMax, where a subchannelized uplink can be used to trade-off throughput for range, achieving a higher range or increased building penetration for a given CPE transmitter power.

IEEE 802.16 in Practice

The familiar model of an IEEE standard being taken up by a trade organisation to drive product and market development is also applicable for 802.16. The WiMAX (Worldwide Interoperability for Microwave Access) Forum aims to fulfil a similar role to that of the Wi-Fi Alliance in relation to the 802.11 suite of standards, namely conformance and interoperability testing and certification.

The WiMAX Forum is marketing 802.16d networks under the WiMAX brand as providing fixed, portable and ultimately mobile wireless broadband access without requiring direct line-of-sight to a base station. Data capacities of up to 40 Mbps per channel are headlined, sufficient to support thousands of residential subscribers at DSL connection speeds, while mobile networks are expected to reach 15 Mbps.

Initial certification work by the WiMAX Forum members will focus on equipment operating in the 3.5 GHz frequency band with 3.5 MHz channel width, and based on both TDD and FDD. Extensions beyond this initial scope will be dependent on further market demand and product submissions by vendors.

Future Developments

Further developments of the 802.16 standards suite are currently being progressed by the TGf and TGg Task Groups. 802.16f aims to improve the multi-hop capabilities, as SSs move between BSs in a fixed wireless network, while 802.16g will deliver faster and more efficient handover and improved QoS for mobile connections.

Other WMAN Standards

ETSI HIPERMAN

HIPERMAN, which stands for high performance radio metropolitan area networks, is a proposal from the European Telecommunications Standards Institute (ETSI) to provide "last mile" fixed wireless access to the small and medium enterprise and residential markets, with the primary aim of accelerating the uptake of broadband Internet access within Europe.

The standard has been developed in close co-operation with the IEEE 802.16 Working Group, and is intended to be interoperable with a subset of the 802.16a standard — namely the OFDM air interface described in

Table 12-9: HIPERMAN Key Parameters

Parameter	*HIPERMAN standard*
RF band	5.725–5.875 GHz (European Band C)
EIRP	< 30 dBm (1 watt)
Data rate	Peak data rates > 2Mbps
Modulation	OFDM (BPSK, QPSK, QAM)
Channelisation	5 MHz, 10 MHz or 20 MHz channel widths
Duplex method	TDD and FDD are supported

the Section "802.11 PHY Layer, p. 310". Table 12-9 shows the key parameters of the HIPERMAN standard.

ETSI's intent is to develop a standard which uses unlicensed spectrum, but which can support the commercial provision of fixed wireless access (FWA) based on either point-to-multipoint (PMP) or mesh network configurations. The channel definition is intended to enable at least two competing service providers to cover a metropolitan area, more if the narrower (5 MHz or 10 MHz) channels are used.

HIPERMAN defines only the PHY and Link Control layers and, as for 802.16d and the WiMAX Forum, it is anticipated that specifications for the Network layer and higher layers will be developed by other bodies. The equivalent "HIPERMAN Forum" is yet to emerge, but the WiMAX Forum has committed to addressing the issue of interoperability, so that HIPERMAN is incorporated into the WiMAX standard. The WiMAX forum will effectively fulfil the commercialisation role for both wireless MAN standards.

TTA WiBro

WiBro, short for Wireless Broadband, is a wireless MAN standard developed by the Telecommunications Technology Association (TTA) of South Korea, Phase 1 of which was approved in November 2004. The standard was developed to fill the gap between 3G and WLAN standards, providing the data rate, mobility and coverage required to deliver Internet access to mobile clients via handheld devices.

The standard uses 100 MHz of licensed RF spectrum, from 2.30 to 2.40 GHz, allocated by the South Korean Ministry of Information

Table 12-10: WiBro Key PHY Layer Parameters

Parameter	WiBro standard
RF band	2.300–2.400 GHz
Network topology	Cellular structure, ca. 1 km range
Maximum data rate; User	Downlink; 6 Mbps, uplink; 1 Mbps
Maximum data rate; Cell	Downlink; 18.4 Mbps, uplink; 6.1 Mbps
Multiple access	OFDMA
Modulation	QPSK, 16-QAM, 64-QAM
Channelisation	9 MHz channel widths
Duplex method	TDD

and Communication for mobile wireless Internet usage, and adjacent to the international unlicensed 2.4 GHz ISM band. The IEEE 802.16-2004 and Draft 3 of the 802.16e standard were the basis for the development of WiBro, and the key PHY layer parameters, shown in Table 12-10, are compatible between the two standards.

The WiBro MAC supports three service levels including guaranteed QoS for delay sensitive applications, based on real time polling of station requirements, and an intermediate QoS level for delay tolerant application that require a minimum guaranteed data rate.

Phase 2 of the standard is planned to focus on network capacity enhancement technology, including MIMO radio, adaptive antenna systems and space time coding, as well as further standardisation with 802.16e-WiMAX.

Metropolitan Area Mesh Networks

The wireless MAN standards described above, enable point-to-point or point-to-multipoint solutions for metropolitan area networking, with dedicated base stations providing centralised control. Although no metropolitan area mesh networking standard is currently under development, the work of 802.11 Task Group TGs described in the Section "Mesh networking (802.11s), p. 167" will blur the boundary between

LAN and MAN scales, allowing 802.11 based mesh networks to operate effectively over metropolitan areas.

A mesh based approach will have a number of advantages over the traditional MAN topologies, including making optimum use of alternative paths through the mesh to maximise network throughput and automatically taking advantage of any new backhaul links that become active in the mesh operating area.

Proprietary (i.e. non standards based) equipment is currently available that operates a pseudo 802.11b mesh, with fixed "mesh routers" providing metropolitan scale coverage for mobile 802.11b devices (see the Section "Wireless LAN Resources by Standard, p. 367").

Summary

Although the WiMAX Forum members are yet to bring IEEE 802.16 compliant products to the marketplace, it looks set to become the *de facto* standard in this sector of the wireless networking market.

The 802.16 MAC enables bandwidth allocation and quality of service to be specified for individual connections, provides adaptive burst profiling to allow the most efficient modulation and coding methods to be used according to each subscriber station's capabilities and environmental conditions, and provides a variety of mechanisms to vary bandwidth allocation according to the requests of individual subscriber stations. Through these features, the 802.16 standard meets the scalability, flexibility and quality of service requirements for metropolitan area networking.

The rapid development of the WiBro standard by the South Korean consumer electronics industry is the first visible chink in the IEEE's armour. Targeted at a specific niche-need in the highly "net-savvy" South Korean consumer market, it shows that speed of development of standards is one area where the IEEE process can fall short of market needs. Nevertheless, WiBro remains based on the 802.16 specifications, stressing the importance of standardisation in global markets.

Chapter 13 now turns to the practical aspects of planning and implementing wireless MANs.

CHAPTER 13

Implementing Wireless MANs

When setting out on the implementation of a wireless MAN project, the starting point will be a specific area envisaged as the general target area of the network. This may be an urban area such as a town or city centre, where potential subscribers will be relatively concentrated, or a more distributed rural area. In the first case the wireless MAN is more likely to be in competition with other broadband access options, such as cable or xDSL, while, in the rural setting, wireless may be the only "last mile" access option available to prospective subscribers.

Two aspects of planning need to be addressed in implementing wireless MANs — technical planning and business planning. The first looks at what is required to build an effective MAN from the technical standpoint — the physical hardware, its specification, location, installation and operation, while the second looks at what is required to operate the MAN as a successful and profitable business. The most important factors to ensure success in this area are an understanding of the market, insight into the competitive situation, and well thought out business and financial plans.

Technical Planning

Technical planning starts with an understanding of the customer demographics in the intended MAN operating area. Based on this understanding, together with a physical and RF survey of the target area, an equipment plan can be developed to achieve maximum effective coverage at minimum capital cost. Future growth of the MAN will be eased by also considering a cost-effective migration path in the planning at this stage — even if this may seem premature.

Site Surveying

Site surveying for a wireless MAN aims to assess the physical and RF environment in which potential subscriber stations will have to operate, for example in terms of physical obstacles and potential sources of interference.

A general site survey should be conducted over the whole target area at the start of the technical planning of the network, to assess the main constraints and consideration that will impact on the overall network design. Later, in the start-up and operating phase, specific surveys of subscriber sites will often be required in order to ensure that the required quality of service can be delivered before commencing physical installation.

A variety of simulation programs, often based on the AT&T Wireless model and more recently the Stanford University Interim (SUI) models, can be found online that can assist in the survey process (see the Section "Wireless MAN Resources by Standard, p. 369". These tools can be used to provide an initial coverage estimate, but they should be used with caution as the performance of the network will be largely determined by local environmental conditions that can only be properly accounted for by on-site physical and RF surveys.

The results of the physical and RF surveys will provide essential input to network planning, determining the best network equipment specification and configuration to avoid potential sources of interference as well as line-of-sight obstacles, while allowing for other specific local conditions.

As discussed below, antenna selection is perhaps the most important element in equipment planning, and the suitability of the selected base station and subscriber antennas will be a key factor in determining the ultimate performance of the network. This selection will be based in part on the site survey results.

Physical Site Survey Considerations

A physical site survey will check the visual line-of-sight from various points in the target area to the potential base station locations. Fresnel zone clearance (see the Section "Fresnel Zone Theory, p. 113") should also be considered as well as the direct line-of-sight (Figure 13-1).

The survey should consider local topography, possible obstructions such as tall buildings, and the proximity to locations such as airports where radar may be a potential cause of interference (this will be checked in the RF survey).

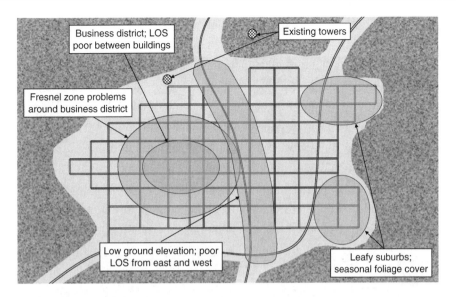

Business district; LOS poor between buildings

Existing towers

Fresnel zone problems around business district

Low ground elevation; poor LOS from east and west

Leafy suburbs; seasonal foliage cover

Figure 13-1: A Physical Site-Survey Map for a MAN Installation

At the subscriber set-up stage, the physical site survey can include specific customer premises equipment (CPE) siting, as well as cable routing and other needs such as lightning protection in exposed locations.

RF Site Survey

This part of the site survey will assess the RF environment in which subscriber stations will have to operate in terms of potential sources of noise and interference. With this objective in mind, the survey should ideally be conducted with an antenna similar to the CPE equipment that will be installed on the subscriber's premises.

An RF site survey is conducted using a spectrum analyser, which will identify wireless transmissions in the target area and in the frequency range of interest. An example of a spectrum analyser display, showing vector analysis of an OFDM signal, is shown in Figure 13-2. The signal strength, direction and polarisation of any signals strong enough to interfere with the network should be recorded, as well as the noise floor (see the Section "Receiver Noise Floor, p. 110").

Spectrum analysis software is also available to run on a desktop or laptop computer and will analyse the signal received from a PCI or PC card receiver equipped with an external antenna.

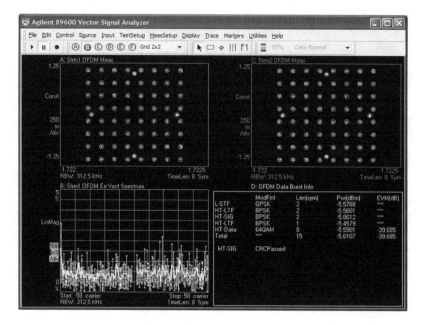

Figure 13-2: Spectrum Analysis Software Display From Agilent
(Courtesy of Agilent Technologies Inc.)

The results of the RF site survey will be important considerations in the technical design of the MAN, and may impact on aspects such as frequency band selection and base station equipment specifications and location.

Capital Equipment Selection and Location

Having completed both physical and RF site surveys, and established potential base station locations to serve the target area, a link budget can be calculated for typical and extreme subscriber locations within the target operating area.

This will establish the transmitter power and antenna gain requirements at each base station as well as the CPE antenna gain required to achieve the desired system performance, taking account of any regulatory limitations (such as EIRP) on the type of equipment that can be used.

Four key elements need to be considered in selecting and locating capital equipment;

- Base station transmitter/receiver
- Base station antenna

- Base station antenna location

- Customer premises equipment.

Selection of a Base Station Transmitter/Receiver

The base station transmitter/receiver (Figure 13-3) is key to achieving maximum effective coverage of the target area. Besides transmitter power and receiver sensitivity, quality and reliability are the factors which will determine overall system performance and, by keeping maintenance and downtime to a minimum, will also reduce operating costs and assure subscriber satisfaction.

Base Station Antenna Selection

Base station antennas come in many shapes and sizes but in most wireless MAN applications omnidirectional coverage will be required. This may be achieved with either omnidirectional or multiple sector antennas (Figure 13-4).

Directional antennas may also be required, for example to extend a MAN using a point-to-point bridge.

In general, within local regulatory limitations, it is preferable to provide the maximum possible antenna gain at the base station, since, for a given link budget, this allows a reduction in the specification and cost of CPE. Installing higher cost antennas at a few base stations is clearly preferred

Figure 13-3: A Base Station Transmitter (Courtesy of Aperto Networks Inc.)

Figure 13-4: A Base Station Sector Antenna Array
(Courtesy of European Antennas Ltd.)

from an economic standpoint to providing the same increment in gain at many more subscriber locations.

Base Station Antenna Location
Having mapped out the intended area for the MAN, the local terrain characteristics and availability of suitable antenna locations would have been assessed as part of the physical site survey. The results of this survey will be analysed to define the minimum number and optimal location of base stations required to serve the target area.

There are two basic options to consider in deciding how to achieve coverage of the target area — coverage from the periphery (Figure 13-5) or from the centre as shown in Table 13-1.

If IEEE 802.16 is the MAN technology being deployed, an individual base station range of 3–10 kilometres should be considered when mapping

Table 13-1: MAN Target Area Coverage Options

Option	*Considerations*
Coverage from the periphery of the target area	A number of locations are identified to provide coverage from the periphery of the target area. This option may be advantageous where local terrain, such as nearby hills, can provide line-of-sight coverage of a wide area. This may also allow subscribers to select the best base station from several different directions.
Coverage from the centre of the target area	In this option one or more base stations are placed on tall structures located within, and preferably close to the centre of, the target area.

out coverage. Depending on the size and shape of the target area, a combination of peripheral and central coverage may be required (Figure 13-6).

Either of these options can be supplemented by the use of additional bridge linked base stations that can be positioned to fill in parts of the target area that may be shielded by high buildings, local terrain or other obstacles.

Having established the general layout of base stations required to cover the target area, Table 13-2 summarises issues that need to be considered in determining specific locations for BS antennas.

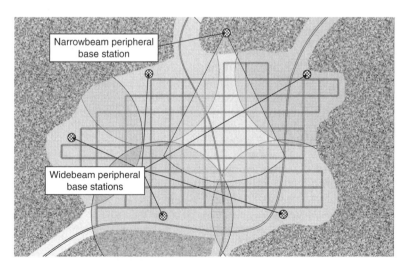

Figure 13-5: Peripheral Coverage of a MAN Target Area

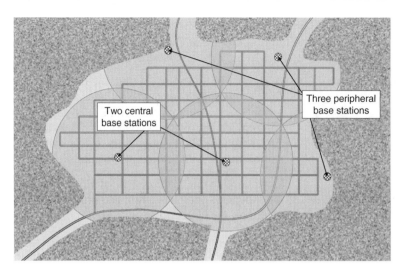

Figure 13-6: Mixed Central and Peripheral Coverage of a MAN Target Area

Height is generally the greatest asset when locating BS antennas, although care must be taken to ensure that the antenna's beam pattern is directed to provide coverage close to the base of the tower or building, for example by rotating the axis of the antenna downwards below the horizon.

Table 13-2: MAN BS Antenna Location Considerations

Issue	*Impact*
Existing buildings or towers	Existing buildings or towers provide a potentially low cost solution for BS antenna location. Barter deals (trading broadband access for antenna siting rights) can also reduce the cost of leasing antenna mounting space on a tower.
New towers	New towers allow optimum location — improved coverage may offset the additional cost compared to leasing space on existing structures. Local planning regulations will need to be considered in siting new towers.
Local terrain	Favourable ground elevation can help areal coverage by adding effective height to an antenna location — whether on the periphery or central to the target area.

Customer Premises Equipment

A typical customer premises equipment (CPE) installation will comprise an antenna and radio enclosed in a weatherproof sealed unit, typically mounted on a suitably facing wall of the premises or on a chimney breast (Figure 13-7). The antenna may range from a patch antenna, with 8–14 dB gain for short range links, to a high gain parabolic reflector antenna with 24 dB gain and upwards for longer range.

Transmitter power will depend on local regulatory limits with 100–200 mW being typical. Cabling will require outdoor rated Cat 5 Ethernet cable as well as a low voltage DC power line routed back inside the premises.

Two approaches can be taken in selecting CPE — one-size-fits-all or subscriber customisation as shown in Table 13-3.

A practical alternative is to start with a one-size-fits-all approach, targeting subscribers close to the base stations, and then to introduce a limited number of higher specification CPE configurations as the network extends.

Table 13-3: Alternative CPE Selection Strategies

Issue	*One size fits all*	*Customisation*
CPE type (antenna, pre-amplification, receiver)	Same CPE selected for all subscribers (e.g. a 14dB patch antenna)	CPE selected according to specific subscriber site survey (e.g. high gain antenna at the limit of MAN coverage)
Coverage	Limited by selected CPE capabilities	Maximised by CPE customisation
Installation complexity and costs	Low	High, requires individual site survey to select the optimal CPE
Capital costs	Lower average cost per subscriber	Higher average cost per subscriber
Subscriber satisfaction	At risk when the limits of standard CPE performance are reached	Likely to be higher as a result of higher level of perceived service through customised installation

Figure 13-7: A Typical CPE Installation from Aperto
(Courtesy of Aperto Networks Inc.)

The cost of installing CPE can be a significant part of the total set-up cost for each new subscriber, and ease of installation is therefore an important consideration in deciding what equipment to use.

Backhaul Provision

Backhaul facilities (Figure 13-8) will provide the link that connects the network base station through to an Internet gateway — the first onward destination of subscriber traffic. This link will prove easiest to achieve if there are local ISPs or other Internet points-of-presence (POPs) nearby.

If a local POP is not available it will be necessary to consider local leased options such as cable or fibre optic providers. In the absence of existing backhaul infrastructure, which may be the situation in remote rural locations, wireless links such as long range point-to-point or satellite options can be investigated.

Business Planning

While technical planning is the key to the physical performance of a MAN, business planning is the key to turning that technical success into a

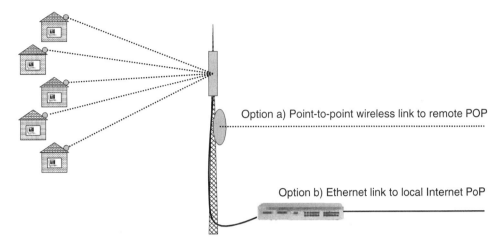

Option a) Point-to-point wireless link to remote POP

Option b) Ethernet link to local Internet PoP

Figure 13-8: Backhaul Configurations

financial success. Although creating a business plan may seem to be a time-consuming effort, that effort will be quickly rewarded by helping to identify any shortcomings that can be more easily, and generally more cheaply, overcome at an early stage in the venture.

The four key elements of the business plan are;

- A description of the business

- A marketing plan

- A management and operations plan and

- A financial plan.

Business Description
This section of the business plan will provide a simple and clear description of the planned business, setting out its purpose and the nature of the service being offered — in this case broadband wireless access. A brief description of the target market, and the specific needs of the market that the service will address, will provide an introduction to the marketing plan.

Marketing Plan
The marketing plan starts with competitor analysis and customer mapping. The competitor analysis will identify alternative broadband access options that are available in the target area, such as cable, xDSL or

competing wireless providers. Some specific aspects of the competition that will need to be assessed are;

- What types of services are being offered by competitors?

- What is the market perception of the quality of service provided?

- What is the range of available uplink and downlink data rates?

- What are the initial set-up and equipment charges made by different competitors?

- What are the typical subscriber charges, including variable rates based on time or usage? Are promotional rates or discounts being offered?

Customer mapping establishes the density of potential subscribers in the target area of the network, including the mix of business and residential customers. This will also be an input to the initial technical planning to generate a picture of the physical network deployment required to reach the maximum number of potential subscribers. To turn the customer mapping into a subscriber forecast (Figure 13-9) it will be necessary to do some market research in order to estimate an uptake rate within the target customer base.

An initial pricing assumption will be needed as input to this market research. This may be based on information from the competitor analysis or on the cost of similar services in other areas. A useful approach may be to

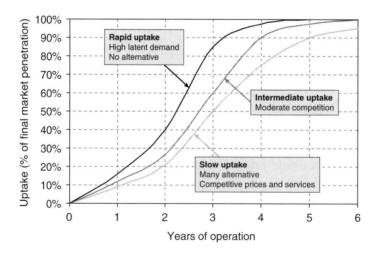

Figure 13-9: Example of Uptake Curves for Various Assumptions

test the likely customer uptake against a range of possible prices as this will allow a range of scenarios to be developed at the financial planning stage.

However, price is not the only way to differentiate a new service from the competition. The overall service bundle can be an important differentiator in the eyes of the subscriber, so it is important to consider creative ways to add distinctive value through additional services that can be delivered at low cost to the service provider. Examples may be web hosting, anti-spam or anti-virus screening or VoIP services.

In considering these aspects of the marketing plan it is important to understand;

- What subscriber needs does the service fulfil and what related needs might be fulfilled?

- What is unique about the service being offering? Bundled services? Lowest cost?

- What is the intended subscriber view of the service? Low cost? Premium service?

- How do competitors position themselves in relation to these service and price aspects?

The final aspect of the marketing plan is advertising — what approaches will be used and what will the cost of advertising be, again as input to the financial plan.

Management and Operations Plan

The management plan will set out how the venture will be managed; who will form the management team, what are their specific qualifications, skills and relevant experience.

An assessment of the strengths and weaknesses of those involved in setting up the venture is important here; What specific expertise and qualifications do they bring to the venture? Where will the technical, business and operational skills needed to run the venture come from? Recognising areas where skills need to be supplemented, is a good starting point for selecting partners and staff.

The operations plan will either be included here or as a separate section. This will cover the additional personnel requirements in the operating phase, as well as outline operating procedures that will need to be developed later.

Other issues such as business insurance can also be addressed in this section of the plan.

Financial Plan

The financial plan should start with a statement of the financial objectives of the venture, in terms of a target revenue or net profit level.

The technical plan, generated in parallel with the marketing plan, will allow capital equipment lists and the resulting start-up budget to be prepared. Similarly monthly and annual operating budgets can be generated from supplies lists, manpower costs, etc. (Table 13-4).

The results of the marketing plan and operating budget can be used to develop a number of cash flow projections, taking account of capital

Table 13-4: Operating Cost Elements and Assumptions

Operating cost item	*Business case assumption*	*Comments*
Base station lease costs	Negotiated costs per month per base station	Covers the space required for equipment located indoors as well as antenna site lease cost
Equipment surveillance and maintenance	Percentage (typically 5%) of BS equipment cost, or 7% of CPE cost if owned by the network operator	Maintenance and costs will be higher for equipment which is remotely located
Network operations	Percentage (typically 10%) of gross revenue in first year, dropping (to 5–7%) in later years	Initial higher % reflects fixed start-up costs, later years reflect stable business position
Sales and marketing costs (including customer support)	Percentage (typically 20%) of gross revenue in first year, dropping (to around 10%) after five years	As for network operations
General & administrative costs	Percentage (typically 5–6%) of gross revenue in first year, dropping (to around 3%) after five years	As for network operations

	May	Jun	Jul	Aug	Sep	Oct
Subscribers (Uptake curve b)	15.0	30.0	50.0	75.0	105.0	140.0
Monthly subscription	10.0	10.0	10.0	10.0	10.0	10.0
Gross revenue	**150.0**	**300.0**	**500.0**	**750.0**	**1050.0**	**1400.0**
Base station leases	150.0	300.0	450.0	600.0	600.0	600.0
Equipment surveillance and maintenance	50.0	100.0	150.0	200.0	200.0	200.0
Network operations	50.0	50.0	50.0	75.0	105.0	140.0
Sales and marketing	100.0	100.0	100.0	150.0	210.0	280.0
General and administration	30.0	30.0	30.0	45.0	63.0	84.0
Total operating costs	**380.0**	**580.0**	**780.0**	**1070.0**	**1178.0**	**1304.0**
Net cash flow	**−230.0**	**−280.0**	**−280.0**	**−320.0**	**−128.0**	**96.0**

Figure 13-10: A 6-Month Initial Cash Flow Projection for a Start-up WISP Venture

spending, direct and indirect expenses and revenues. These should be made on a monthly basis for the start-up period of one to two years and then on a yearly basis. An example start-up cash flow projection is shown in Figure 13-10.

Generating multiple scenarios based on alternative uptake rates and pricing policies will illustrate the range of potential cash flow outcomes and help to assess how exposed the venture is to risk. Cash flows can be used to identify the total funding required until the venture starts to be positive cash generating.

Cash flows can then be turned into income projections (profit & loss statements) and various break-even and profitability analyses can be performed (Figure 13-11). Again, multiple scenarios on customer uptake, pricing, future market development and other variables can be played out to give stakeholders an idea of the robustness of the proposed venture.

Spreadsheets are an ideal tool for conducting this analysis, although custom made financial and general business planning software is also available that can guide the financial planning process.

Start-up Phase

Some of the key considerations that will require attention in the start-up phase are covered in this section.

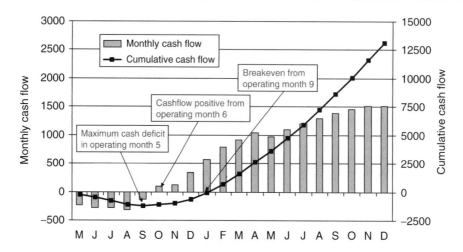

Figure 13-11: A Graphical Break-even Analysis Chart

Base Station Deployment

Tower leasing — Where existing towers are the preferred option, contracts will have to be negotiated to acquire space to locate BS antennas. Sample tower leasing agreements can be found on the Internet, although it is recommended to take legal advice to ensure that agreements comply with local laws.

New tower deployment — When selecting a tower design, current needs and requirements for future expansion should be considered. Construction may be subject to local planning conditions such as improving site access, providing an equipment shed and lighting, etc. Planning authorities may also require a waiver to allow future use of a new tower by other operators at a reasonable rent. Soil sampling may be required depending on the design of tower footings — the structural engineer involved in tower design will need to advise on this requirement.

Physical BS antenna deployment — Positioning antennas, waterproof cables and securing all connections with the aim of minimising future maintenance costs.

Subscriber Deployment

Subscriber agreement — A subscriber agreement will be required to define the terms and conditions of the service being provided. Besides the obvious aspects such as cost and the duration of the contract, the agreement should cover any quality of service guarantees, liabilities, termination conditions, etc. An Internet search will turn up a wide variety of sample subscriber agreements, including many for wireless ISPs, although once again it is recommended to take legal advice to ensure that agreements comply with local laws.

Subscriber site survey — Once the BS is installed, a simple site survey can be conducted at each new subscriber site by using the intended CPE antenna and receiver mounted on a light pole to lift the antenna to the intended installation point, with a cable length comparable to that required to run back to the subscribers computer location. A preliminary link budget calculation, followed by this type of quick site survey, can assure the quality of the wireless link and help prevent repeat visits in the event of performance problems.

Customer premises equipment installation — If ease and speed of installation have been considered when selecting equipment, CPE installation can be prevented from consuming too much time and money as the network grows.

Customer premises equipment grounding — Effective CPE grounding is important for three reasons, to make sure the antenna is operating efficiently, to comply with any applicable local electrical installation regulations and for lightning protection. Connecting to the building's ground conduit will usually be sufficient, but local building regulations should be consulted to ensure compliance.

Operating Phase

Some of the key considerations that will require attention in the operating phase are described below.

Technical Operations

Customer helpline — The low cost start-up option is to use an answering service to help customers with basic instructions, FAQs and to point them to other more technically oriented help resources. If subscribers need support on more general networking/computing issues, an advanced technical support service could be provided but this may consume a significant slice of monthly revenues.

Subcontracting CPE and BS installation — Operator self-installation of CPE is the low-cost option for start-up, but a trained team of part-time installers, who are able to work flexible hours, will be an ideal solution once the venture is up and running. Scheduling and managing the team is a task that can be automated using e-mail and a suitable scheduling system.

Business and Financial Operations

Subscriber billing — Many off-the-shelf software systems are available for ISP billing. Typically these systems are designed to handle ISP specific features such as variable rates (flat, tiered, time/day or usage based), free offers (hours or usage), pre-paid card support and e-mail reports, reminders or invoices.

Incentives — Can be an effective way to attract new subscribers. Existing subscribers can be given incentives to refer new customers, and others, such as installation contractors, can also be encouraged to promote the business by the incentive of extra work.

Managing network operating costs — Controlling ongoing costs and achieving revenue targets will be a key focus in the operational phase. A clear understanding of the make-up of total network operations costs will be important, covering elements such as installation, infrastructure surveillance, maintenance, backhaul and administration costs.

Business accounts — Options include DIY, with many software systems available, or hiring a part-time accountant who can either train on an existing accounting system or set one up for the venture. A small-scale WMAN venture should require no more than a day or two a month to manage all the business accounts.

Summary of Part V

Wireless metropolitan area networking is set for future growth following the publication of the initial IEEE 802.16 suite of standards and the progress towards mobility and mesh networking under development by Task Groups TGe, TGf and TGg.

These standards provide the essential networking capabilities required in WMAN applications, scalability, service flexibility and quality of service — capabilities that are beyond simpler MACs such as that specified in the original IEEE 802.11 standard.

Standard compliance and interoperability is being progressed by the WiMAX Forum, and as certified products start to emerge, the promise of ubiquitous wireless broadband access that motivated the original IEEE 802.16 development will begin to be realised.

PART VI

THE FUTURE OF WIRELESS NETWORKING TECHNOLOGY

Introduction

Besides showing the current status of wireless networking technologies, Parts III to IV have highlighted the areas where, over the full wireless range from a few centimetres to many kilometres, the further development and enhancement of these technologies is continuing. Aspects such as quality of service, roaming and satisfying the ever increasing demand for data bandwidth are the focus areas of many current developments.

In this chapter some key developments are outlined that go beyond the incremental enhancement of existing wireless networking concepts. These developments, such as cognitive radio and media independent handoff, typically bridge across distance scales and are likely to be significant drivers of change in the fundamental nature of wireless networking in the coming years.

CHAPTER **14**

Leading Edge Wireless Networking Technologies

In this chapter four key technologies are discussed that are currently under development and that will play a large role in shaping the future of wireless networking;

- Wireless mesh network routing

- Network independent roaming

- Gigabit wireless LANs

- Cognitive, or spectrum agile radio.

While individually these are significant step-outs from existing technologies (for example, cognitive radio radically extends the 802.11h enhancements to 802.11a networks that were described in the Section "Spectrum Management at 5 GHz (802.11h), p. 160"), together they herald a not-too-distant future in which spectrum availability, propagation range and data bandwidth are no longer limiting factors for wireless network performance.

Wireless Mesh Network Routing

As briefly described in the Section "Mesh Networks, p. 43", wireless mesh networks or mobile ad-hoc networks (MANETs) offer some significant advantages for large-scale wireless networking, including;

- self-organising architecture, optimising routing and traffic distribution

- self-healing ability to respond to broken or unreliable wireless links

- increasing network throughput as the density of devices increases.

A major challenge faced in defining mesh networking standards is to design Data Link layer protocols that are able to achieve this flexibility without consuming an excessive amount of network bandwidth for routing and control messaging. Simple, low overhead approaches to this problem, such as making routing decisions based on RF signal strength or the minimum hop count from source to destination, do not perform well compared to routing algorithms that actively probe the mesh topology and make routing decisions based on the historical and predicted throughput of the available paths through the mesh.

One intriguing approach being investigated for mesh network routing is inspired by the communication method that enables ants to converge on an optimum route to food sources, while maintaining back-up routes that can be used in the case of overcrowding or other obstacles.

Ants use a method of communication called stigmergy, in which each ant modifies its local environment by laying down a chemical trail of pheromones and other ants respond to these modifications in such a way that the global behaviour of the colony becomes coordinated. Because the pheromone is volatile, and the intensity decays naturally with time, short, fast and often used routes will have a higher pheromone intensity than long, slow, blocked or abandoned routes, and will therefore be more often used and reinforced. Adaptation and improvement of existing routes, as well as the discovery of new routes, arises as a result of a degree of randomness inherent in the process.

Some key elements of MANET routing algorithms inspired by this biological system are summarised in Table 14-1.

While some ant colony inspired routing algorithms use either reactive or proactive strategies alone to gather information, combining these two approaches, and adding stochastic routing, results in a system that more closely mimics biological ant behaviour.

It remains a challenge to minimise the bandwidth overhead used by the route sensing ant-agents, but algorithms that explicitly imitate aspects of ant behaviour may prove to be a key enabler for large-scale mesh networking.

Table 14-1: Features of Ant Colony Inspired MANET Routing Algorithms

Routing feature	*Description*
Pheromone tables	A table of routing information maintained in each node of the mesh, which indicates the "goodness" of the link to each of its neighbours in terms of data packet delivery time and number of hops to a destination.
Reactive routing information gathering	Software agents, unsurprisingly called ants, are generated to update pheromone tables in response to events such as a new station joining the mesh or the failure of a previously reliable route. Typically a "forward ant" seeks out a route from source to destination and a "backward ant" returns over this route updating the tables in each intermediate node.
Proactive routing information gathering	Ants are periodically generated to proactively sample and optimise existing routes as well as discover alternative routes. Pheromone tables are updated to continuously optimise and evolve routing decisions as well as to respond to disruptive events.
Stochastic routing decisions	When several alternative paths are available for a data packet's next hop, a path is selected stochastically, giving higher probability to the path with the highest pheromone table value. This leads to automatic load balancing, since data is distributed across all good paths, and if one becomes overloaded it will be avoided until the congestion eases.

Network Independent Roaming

Media Independent Handoff

In the Section "Network Performance and Roaming (802.11k and 802.11r), p. 162", three situations were described where client stations need to make transitions between WLAN access points; for mobile client stations moving out of range of a current access point, to maintain service availability under changing environmental conditions or service needs, or for load balancing within the WLAN. Transitions between points of attachment (POA) for a single network type (such as an 802.11 WLAN) are termed homogeneous transitions, and in the case of 802.11 networks, Task Groups TGk and TGr are developing and enhancing the mechanisms that enable seamless WLAN transitions.

Table 14-2: Roaming Needs Requiring Heterogeneous Transitions

Roaming need	Description
Mobile client; coverage	A client station may move out of range of its current POA, and need to transition to another network type because the current type is no longer available; e.g. moving out of range of an 802.11 hotspot and transitioning to a cellular phone service to maintain a voice connection.
Mobile client; cost advantage	A mobile client may move into range of an alternative network that is able to provide the same or better QoS as the current POA but at a lower price; e.g. transitioning a voice call from a cellular phone service to a VoIP service when moving into range of an 802.11 hotspot.
New service requirement	A new application is started that requires a level of service that is not supported by the current POA; e.g. downloading a large file may be able to take advantage of a higher data rate available on another network.

The next step in providing uninterrupted connectivity to the mobile user is to be able to make similar seamless transitions across multiple wireless networks of different types. These so-called heterogeneous transitions might involve a single user connection successively handing-off from an 802.11 WLAN to a cellular phone service and then to a WiMAX MAN and finally back to a WLAN via a new access point. As for homogeneous transitions, there are a number of reasons why mobile users may want to make this type of heterogeneous transition, as shown in Table 14-2.

The challenge of enabling uninterrupted, QoS guaranteed, heterogeneous hand-offs is being addressed in the IEEE 802.21 Working Group, which started work in March 2004. 802.21 defines media independent handover (MIH) mechanisms that enable networks such as Wi-Fi, WiMAX and cellular phone networks to co-operate at the Network and Data Link layers of the OSI protocol stack (see Figure 14-1).

The MIH function is a unified technology-independent interface that provides inputs and context to the upper layers to assist in handover decision-making. In turn the MIH gathers the necessary information on link parameters such as uplink /downlink rate, signal strength and range, and link capabilities such as QoS and security. This information is

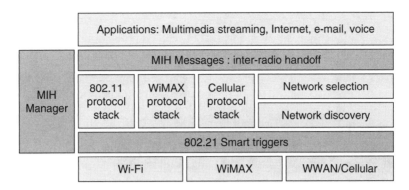

Figure 14-1: MIH Function in the Protocol Stack of a Multi-Radio Mobile Device

gathered through technology-specific Layer 2 service access points for each of the enabled technologies, such as 802.11, 802.16 and 3GPP/3GPP2 for cellular phone networks.

MIH in Practice

The first devices and service that are putting MIH into practice are targeting the personal telephony sector, enabling cellular phone subscribers to use a single device to access cellular services when on the move and VoIP services when at home.

BT's Fusion service, launched in the UK in 2005, uses Bluetooth enabled Motorola handsets together with a Bluetooth hub as a gateway for VoIP calls connected via a BT ADSL broadband Internet connection. Launched in partnership with Vodaphone, this service allows calls to be handed off between the cellular network and the VoIP over Bluetooth connection when the handset comes within range of the Bluetooth hub.

Motorola announced a family of products in early 2006 that enable handoff between cellular phone and 802.11 based VoIP services. The residential seamless mobility gateway (RSG — Figure 14-2) includes an 802.11b/g access point, a four-port router and a VoIP adapter, allowing seamless handoff of voice calls between the home WLAN and the cellular network when using a dual-mode handset.

These devices provide the expected VoIP features, such as 802.11i security and voice traffic prioritisation on the WLAN to ensure QoS,

Figure 14-2: Motorola Residential Seamless Mobility Gateway
(Courtesy of Motorola Corporation)

as well as offering many digital phone features, such as supporting multiple lines, caller ID, call waiting and call forwarding services.

Ahead of the full development and ratification of the 802.21 standard, these devices rely on proprietary software and protocols to achieve a limited degree of media independent handoff and, in the BT case, the handoff only works with specific service providers. Nevertheless, these early demonstrations of the concept provide a foretaste of the remarkable flexibility that full MIH will provide.

Gigabit Wireless LANs

The Section "MIMO and data rates to 600 Mbps (802.11n), p. 165" described how the 802.11 Task Group TGn is working towards delivering a PHY layer data rate of 500–600 Mbps, and an effective MAC SAP rate of 100 Mbps, through modifications to the 802.11 PHY and MAC layers and the application of MIMO radio. A number of standards based and proprietary equipment development projects are also underway aiming to deliver a PHY layer data rate of 1 Gbps, a technology threshold becoming known as Gi-Fi that is motivated by a range of home, office and public usage scenarios (Table 14-3).

One such effort is the WIGWAM project, which stands for wireless gigabit with advanced multimedia, and was initiated in 2003 by a group of European companies and academic institutions to develop the enabling technologies for gigabit WLANs. The WIGWAM project is industry-driven rather than standards-driven, although the consortium intends to present its results to the relevant standards organisations.

Table 14-3: Gigabit Wireless Usage Scenarios

Usage scenario	Description
Home usage	Multiple concurrent high bandwidth media streaming applications (Video and HDTV) requiring data rates of several 100 Mbps per user. Fast synchronisation of personal storage devices exceeding 100 GB capacity.
Office usage	Replacing wired Ethernet in supporting high bandwidth office applications such as high-quality video conferencing, streaming media and network file sharing, with the required security and QoS.
Public access usage	Short range, very high data rate complement to existing public access networks such as GSM, GPRS, Wi-Fi, WiMAX, with seamless media independent handover.
High speed mobile usage	Multi-user broadband Internet and media streaming to cars and trains, with the additional technical challenge of varying Doppler shifts.

The technical challenges faced in reaching this next step in wireless network throughput are familiar from earlier parts of the book;

- maximising spectral efficiency to get more data bits into each hertz of RF spectrum

- maximising MAC efficiency so that most of the transmitted bits are upper layer data rather than Link Control and MAC layer overhead

- ensuring security through strong encryption algorithms, with fast computation on low-cost hardware.

In addressing these challenges, the WIGWAM consortium is considering broadly similar approaches to those under discussion by 802.11 TGn, for example, MIMO radio and OFDM with higher efficiency coding (LDPC and Turbo Codes — see next section) are under consideration to achieve high spectral efficiency. One additional technology being considered by WIGWAM is OFDMA with multi-carrier code division multiple access (OFDMA/MC-CDMA), described in the Section "Multi-Channel Code Division Multiple Access (MC-CDMA), p. 353".

WIGWAM is initially targeting operation at 5 GHz but extensions in the 17, 24, 38 and 60 GHz RF bands are also under consideration. This would bring wireless networking into the millimetre wave bands, with wavelengths of 8 mm at 38 GHz to 5 mm at 60 GHz.

LDPC and Turbo Codes

Low density parity check codes are error correction codes that are computationally efficient and perform close to the theoretical limit in enabling error recovery in noisy communication channels. Unlike a standard parity check, which simply flags bit errors during transmission, these codes also allow errors to be probabilistically recovered i.e. from a set of possible transmitted data blocks it is possible to compute which is most likely to have resulted in the received data block and check code.

A check code is computed from a data block by sparsely sampling bits in the block using a random sampling matrix. Figure 14-3 illustrates a sampling matrix and the 4-bit check code resulting from the example 8-bit input work. The "low density" in LDPC refers to the fact that there are few 1's in the matrix.

In this example, if the input word is incorrectly received as 00100101, the check code word can be used to confirm that the most significant bit should have been a 1.

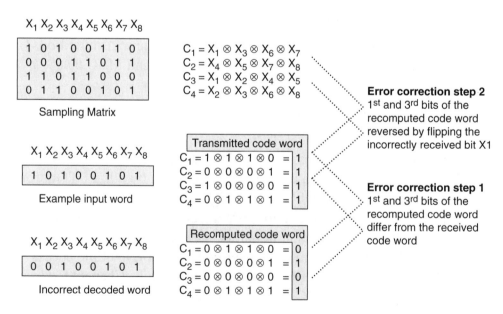

Figure 14-3: LDPC Computation and Error Correction

Turbo codes are another form of error correction code in which two parity checks are performed on the data block, one on the straight data and one on a known permutation of the data. In the receiver, two decoders use the two parity check blocks to compute the most likely sequence of transmitted bits. If the two decoders come up with different results, they exchange information and re-compute the most likely transmitted bit sequence, iterating until the two results are identical.

The advantage of turbo codes is that they achieve very effective error recovery while keeping the coding rate close to 1, while the disadvantages are computational complexity and latency — in view of the iterative decoding process.

Multi-Channel Code Division Multiple Access (MC-CDMA)

As described in the Section "Code Division Multiple Access, p. 94", CDMA assigns an orthogonal Walsh–Hadamard code to each receiver and uses this as a chipping code to spread the input data stream. The orthogonality of chipping codes ensures that each receiver is only able to decode symbols encoded with that user's unique code.

In MC-CDMA (Figure 14-4) each chip is transmitted in parallel using the same number of carriers as there are chips in the chipping code. In this case the orthogonality of codes allows multiple reuse of the same set of OFDM subcarriers by several concurrent users.

Schematic block diagrams of an MC-CDMA transmitter and receiver are shown in Figure 14-5. From the left, a series to parallel converter splits the input bit stream into N parallel stream and each of these is further split into M parallel chip streams by the XOR operation with the M chips of the code word (C_1 to C_M). This results in the input bit stream being spread over a total of $N \times M$ subcarriers.

The parallel chip streams of multiple users are XOR'd together and a modulator maps each resulting chip stream onto the amplitude/phase constellation in use, chip by chip for BPSK or in 6-chip symbols for 64-QAM. The $N \times M$ amplitude and phase points drive the inputs of an Inverse FFT and the computed output signal is transmitted after insertion of a guard interval.

At the receiver, after guard interval removal the FFT computes the amplitude and phase of each of the $N \times M$ subcarriers and a demodulator

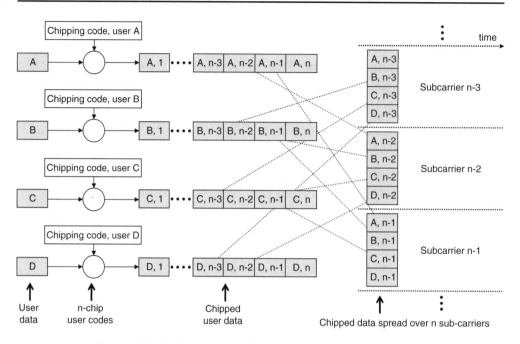

Figure 14-4: Data Spreading in OFDMA/MC-CDMA

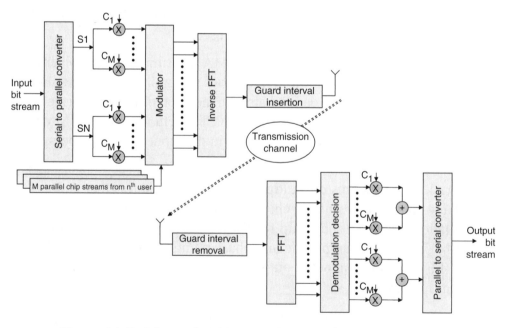

Figure 14-5: Schematic MC-CDMA Transmitter and Receiver

translates these constellation points into the equivalent input chips or multi-chip symbols. The M chips of the receiver's code word are XOR'd with each set of M demodulated chip streams to recover the N parallel bit streams which are finally converted back to a serial bit stream.

Gigabit Wireless in Practice

Siemens AG, a member of the WIGWAM consortium, announced the first 1 Gbps wireless link operating in the 5 GHz band in December 2004, based on a 4×3 (Tx $\times$ Rx) MIMO radio together with an unspecified OFDM method.

By June 2005, a data rate in excess of 10 Gbps was demonstrated by a University of Essex team in the UK, over a 60m line-of-sight link. Three RF bands between 2 and 7 GHz were used to create three concurrent data channels of 1.2, 1.6 and 2.4 Gbps, with each band also supporting a second concurrent channel using polarisation-based frequency reuse.

Broad commercial application of 1 Gbps wireless LANs may be anticipated around 2010.

Cognitive Radio

The concept of a cognitive radio was first introduced by Joseph Miltola and Gerald Maguire in 2000, to represent a wireless device that combines an awareness of the RF environment together with learning and reasoning algorithms that can modify wireless PHY parameters in order to meet user requirements within the constraints of the RF environment.

Spectrum agile radios are similar but with the emphasis on spectrum sensing and adaptation rather than on learning and reasoning. Spectrum sensing devices and algorithms are used to detect other users and enable spectrum agile radios to adjust their transmission parameters in response to the presence of other radios. Spectrum agile radios may also exchange sensing data in order to co-operate in making use of transmission opportunities.

Another key concept underlying cognitive radio is that of a software defined radio (SDR) in which the digital signal processing functions such as data coding and modulation, are performed in software rather than hardware. This gives a cognitive radio the flexibility to use alternative processing schemes, depending on changing requirements.

Radio Frequency Policy Modernisation

In December 2002, an FCC Spectrum Policy Review Task Force concluded with a number of recommendations aimed at modernising the regulatory framework to increase access to the RF spectrum, while assuring the freedom of existing licensed services from interference. This review was motivated by the continually increasing demands on limited spectrum resources, coupled with the observation that, even in those parts of the RF spectrum that are fully allocated, the bandwidth is typically in use only 10–20% of the time.

Following on from this review, the FCC issued a so-called notice of proposed rulemaking (NPRM) in May 2004, proposing to open up the 76–698 MHz portion of the TV spectrum in the US for unlicensed usage. This ruling would allow wireless networks to make unlicensed use of the unused "white space" in these licensed TV broadcast bands. Two types of devices are permitted to make unlicensed use of these bands under this ruling, as shown in Table 14-4.

This opportunity led to the start up of the IEEE 802.22 Working Group in October 2004, which aims to develop MAC and PHY layer specifications for wireless regional area networks (WRAN) to operate in unused VHF/UHF TV bands between 54 and 862 MHz. Spectrum agile radio is the main enabling technology for the 802.22 work.

Table 14-4: FCC Permitted Devices for Unlicensed Operation in TV Bands

Device type	Characteristics
Fixed devices	Maximum transmit power of 1W Must be either professionally installed to operate in locally unused channels or fitted with GPS and a means of determining which channels are locally vacant. This could be an Internet connection to a Spectrum Policy Server which provides location base spectrum coordination.
Mobile devices	Maximum transmit power of 100 mW Must receive a control signal from a device that determines which channels are vacant.

Spectrum Sharing Approaches and Challenges

Approaches to spectrum sharing are broadly categorised as either vertical or horizontal. Vertical, also known as primary, sharing occurs when a channel is identified that is not being used — a so-called "white space" — while horizontal, also known as secondary, sharing is the efficient coexistence of two or more networks simultaneously operating in the same spectrum.

Horizontal sharing relies either on rule based coordination of multiple radios within a single device, such as the Wi-Fi plus Bluetooth coexistence measures described in the Section "Bluetooth Coexistence with 802.11 WLANs, p. 247", or the coordination of radios via a separate common spectrum coordination channel (CSCC) that connects devices using beacon broadcasts in an edge-of-band channel, or by some other means, including potentially via the Internet.

The 802.22 approach is based on vertical sharing, and the key challenge here is spectrum sensing — reliable dynamic detection of the presence of a primary (licensed) user operating in or returning to a channel. Three possible approaches to spectrum sensing are summarised in Table 14-5.

Table 14-5: Spectrum Sensing Approaches for Vertical Spectrum Sharing

Spectrum sensing approach	*Characteristics*
Matched receiver	A receiver matched to the signal characteristics of the primary user (coding, modulation and synchronisation) and with knowledge of other signal characteristics and channel usage (spreading codes, pilots and training sequences, etc.). This provides very reliable detection but is relatively inflexible, although this can be partly overcome using a software defined radio.
Energy detection	Similar to a spectrum analyser, the receiver determines the presence or absence of a primary user by aggregating energy detected in the target channel. This can be effective for narrow band signals, but DSSS, FHSS or other wideband signals may not be detected.
Feature detection	Periodic features of modulated signals are detected by computing a spectral correlation, which detects periodicity from pulse trains, hopping cycles, cyclic prefixes, etc.

Using these sensing approaches, a spectrum agile radio would build up a local spectrum map, identifying the detected power level, signal type (e.g. FM, FHSS, OFDM) and bandwidth in use in each channel, as well as estimating the quality of empty channels in terms of their potential *SNR*.

As an alternative to each spectrum agile radio making decisions based on its local spectrum map, multiple radios could also collaborate in a vertical sharing mode, aggregating channel measurements via a CSCC protocol, in order to identify and exploit "white space" opportunities. If multiple radios compete for transmit opportunities, the CSCC channel would enable the contention to be resolved via a priority system, or alternatively dynamic auctions could be implemented, allowing devices to bid for media access.

A further issue in spectrum sharing is identifying and preserving redundant "stand-by" channels, in case a primary user resumes broadcasting and the spectrum agile radio has to shift channels in order to avoid interfering.

Solving these MAC and PHY layer challenges, starting with the efforts of the 802.22 Working Group in defining the specifications for wireless regional area networks, will enable cognitive and spectrum agile radios to open up the potential of a duty cycle approaching 100% in every hertz of the RF spectrum, unblocking spectrum availability as a constraint for a long time to come.

Summary of Part VI

The four developing technologies briefly described in this chapter promise a not-too-distant future in which wireless networks become truly ubiquitous — with virtually unlimited bandwidth available over a variety of different network types and ranges, enabling always on, mobile connectivity at any desired service level.

The technology challenges of bandwidth, media access, QoS and mobility, have been recurring themes as each scale of wireless networking has been explored in earlier chapters. These new technologies may look set to consign these particular challenges to the history books, but there can be little doubt that new usage models and services will also emerge that will put greater demands on wireless networks and pose new challenges for the standards, software and hardware developers.

It is likely still to be some time before the final chapter can be written on wireless networking technology.

WIRELESS NETWORKING INFORMATION RESOURCES

Introduction

Part VII provides a knowledge base on wireless networking in three sections;

- a quick reference guide to some of the key online information sites and resources, arranged by wireless networking standard

- a comprehensive listing of acronyms commonly used in wired and wireless networking

- a glossary covering some of the key technical terms introduced earlier in the text.

15

Further Sources of Information

The following sections aim to provide a quick reference to some of the key information sites and resources relating to wireless networking, where the latest up-to-date information on the status and further developments of the various standards can be followed.

Any attempt to provide a comprehensive listing of available information on wireless networking in print form is destined to become quickly obsolete and overshadowed by the power of Google. Resources listed here have therefore been selected based on an expectation of longevity and continued relevance.

General Information Sources

Standards organisations	
Institute of Electrical and Electronic Engineers	www.ieee.org/portal/site
IEEE Wireless Standards Zone	standards.ieee.org/wireless
Technology fora	
Ultra Wideband Forum	www.uwbforum.org
The Wireless Association	www.ctia.org
Ultra Wideband Planet	www.ultrawidebandplanet.org
Wireless Communications Association	www.wcai.com
Resources	
RFC archive	www.faqs.org/rfcs

NIST WLAN Security Framework	www.src.nist.gov/pcig/checklists
Wireless Net Design Line	www.wirelessnetdesignline.com
Wireless Design Online	www.wirelessdesignonline.com
Wireless Networking Tutorials	www.wirelessnetworkstutorial.info
Wireless Technology Information	www.radio-electronics.co.uk/info/wireless
Trade publications	
Wireless Week	www.wirelessweek.com
Wi-Fi Net News	www.wifinetnews.com
Wireless News Factor	www.wirelessnewsfactor.com
Mobile Enterprise	www.mobilenterprisemag.com

Wireless PAN Resources by Standard

Bluetooth (IEEE 802.15.1)

Standards group	
802.15 WPAN Working Group	grouper.ieee.org/groups/802/15
802.15 WPAN Task Group TG1	www.ieee802.org/15/pub/TG1.html
802.15 Task Group TG2 (Coexistence)	grouper.ieee.org/groups/802/15/pub
Trade organisations	
Bluetooth Special Interest Group	www.bluetooth.org and www.bluetooth.com
Resources	
Palowireless Bluetooth Resource Centre	www.palowireless.com/bluetooth
The Unofficial Bluetooth Weblog	bluetooth.weblogsinc.com
Blueserker — Berserk About Bluetooth	www.blueserker.com
Bluetooth Shareware	www.bluetoothshareware.com
News Tooth	www.newstooth.com/newstooth
Bluetooth tutorial	www.tutorial-reports.com/wireless/bluetooth
Suppliers	
Directory of Bluetooth products and services	www.thewirelessdirectory.com/Bluetooth.htm

Blueunplugged	www.blueunplugged.com
Ericsson	www.ericsson.com/bluetooth
Nokia	www.nokia.com/bluetooth
Motorola	www.motorola.com/bluetooth

Wireless USB

Standards group	
WiMedia Alliance	www.wimedia.org
UWB Forum	www.uwbforum.org

Trade organisations	
USB Implementers' Forum	www.usb.org

Resources	
Palowireless UWB/Ultra Wideband Resource Centre	www.palowireless.com/uwb
USB-IF WUSB resources	www.usb.org/developers/wusb

Suppliers	
Staccato Communications	www.staccatocommunications.com/products
Belkin	www.belkin.com
Freescale	www.freescale.com
Wisair	www.wisair.com

ZigBee (IEEE 802.15.4)

Standards group	
IEEE 802.15 Working Group	www.ieee802.org/15

Trade organisations	
ZigBee Alliance	www.zigbee.org

Resources	
Palowireless ZigBee Resource Centre	www.palowireless.com/zigbee
Ultrawideband Insider	www.uwbinsider.com
ZigBee tutorial info	www.tutorial-reports.com/wireless/zigbee

Suppliers

Telegesis	www.telegesis.com
Crossbow Technology	www.xbow.com
Freescale	www.freescale.com
Cirronet	www.cirronet.com

IrDA

Standards group

Infrared Data Association	www.irda.org

Trade organisations

IrDA	www.irda.org

Resources

Palowireless IrDA/Infrared Resource Centre	www.palowireless.com/irda
eg3	www.eg3.org/irda.htm

Suppliers

ACTiSYS	www.actisys.com
Clarinet Systems	www.clarinetsys.com

FireWire (IEEE 1394)

Standards group

IEEE 1394 Working Group	grouper.ieee.org/groups/1394/c
IEEE 802.15.3 WPAN Working Group	www.ieee802.org/15/pub/TG3.html

Trade organisations

IEEE 1394 Trade Association	www.1394ta.org

Resources

Palowireless 802.15 WPAN Resource Centre	www.palowireless.com/i802_15
Apple Computer Inc.	developer.apple.com/devicedrivers/firewire

Suppliers

FireWire Depot	www.fwdepot.com/thestore
Global Sources	www.globalsources.com/manufacturers/IEEE-1394-Firewire.html

Near Field Communications (NFC)

Standards group

ECMA	www.ecma-international.org

Trade organisations

NFC Forum	www.nfc-forum.org

Resources

Radio Electronics NFC overview	www.radio-electronics.com/info/wireless/nfc/nfc_overview.php
UNIK RFID tutorial	wiki.unik.no/index.php/Rfidtutorial

Suppliers

Philips	www.semiconductors.philips.com/products/identification/nfc
Nokia	www.nokia.com/nfc
Sony	www.sony.net/Products/felica

Wireless LAN Resources by Standard

Wi-Fi (IEEE 802.11)

Standards group

IEEE 802.11 Working Group	www.ieee802.org/11

Trade organisations

Wi-Fi Alliance	www.wi-fi.org
Wireless LAN Association	www.wlana.org
Enhanced Wireless Consortium	www.enhancedwirelessconsortium.org

Resources

Palowireless 802.11 WLAN Resource Center	www.palowireless.com/i802_11
Wi-Fi Planet	www.wi-fiplanet.com
802.11 News	www.80211anews.com
Wireless Gumph	www.wireless.gumph.org
Wi-Fi tutorial info	www.tutorial-reports.com/wireless/wlanwifi
Homemade LAN antennas	www.wlan.org.uk/antenna-page.html

Suppliers

Proxim	www.proxim.com
Linksys	www.linksys.com
D-Link	www.dlink.com
Belkin	www.belkin.com
Netgear	www.netgear.com
PC based spectrum analyser	www.cognio.com
WLAN analyser	www.netstumbler.com
Bluetooth interference analyser	www.airmagnet.com/products/bluesweep.htm
Site survey and WLAN planning tools	www.wirelessvalley.com

Wireless Mesh (IEEE 802.11s)

Standards group

IEEE 802.11 Working Group	www.ieee802.org/11

Trade organisations

Wi-Mesh	www.wi-mesh.org

Resources

Mobile Pipeline tutorial	www.mobilepipeline.com/howto/21600011
BelAir Networks resources	www.belairnetworks.com/resources

Suppliers (proprietary, pre- 802.11s)

BelAir Networks	www.belairnetworks.com
Nortel Networks	www.nortelnetworks.com
Tropos Networks	www.tropos.com

HiperLAN/2 (ETSI)

Standards group

European Telecommunications Standards Institute	www.etsi.org

Trade organisations

HiperLAN2 Global Forum	www.hiperlan2.com

Resources

Palowireless HiperLAN and HiperLAN/2 Resource Center	www.palowireless.com/hiperlan2

Wireless MAN Resources by Standard

WiMAX (IEEE 802.16)

Standards group

IEEE 802.16 Wireless MAN Working Group	www.wirelessman.org

Trade organisations

WiMAX Forum	www.wimaxforum.org
WiMAX Industry	www.wimax-industry.com

Resources

WiMax.com	www.wimax.com
802.16 News	www.80216news.com
WiMaxxed	www.wimaxxed.com
Palowireless IEEE 802.16 WMAN Resource Center	www.palowireless.com/i802_16
WiMAX tutorial info	www.tutorial-reports.com/wireless/wimax

Starting, operating and maintaining WISPs.	www.startawisp.com
WISP Centric	www.wispcentric.com
Link budget tools	www.wirelessconnections.net

Suppliers

Proxim	www.proxim.com
ACTiSYS	www.actisys.com
Solecktek Corporation	www.solectek.com
Antenna suppliers	www.andrew.com

Cognitive Radio

Standards group

FCC cognitive radio technologies	www.fcc.gov/oet/cognitiveradio

Trade organisations

Software Defined Radio Forum	www.sdrforum.org

Resources

Rutgers cognitive radio resources	www.winlab.rutgers.edu/~xjing/prj/CognitiveRadio.htm
Programmable Wireless	www.programmablewireless.org

Suppliers

Adapt4	www.adapt4.com
GNU software radio	www.gnu.org/software/gnuradio
VANU software radio	www.vanu.com

Glossary

A comprehensive guide to acronyms and other common terms in use in networking and wireless networking.

Networking and Wireless Networking Acronyms

A

AAS	Adaptive (or Advanced) Antenna System
AC	Access Category
ACK	Acknowledge (Flow control frame)
ACL	Asynchronous Connection-Less
AES	Advanced Encryption Standard
AFH	Adaptive Frequency Hopping
AIFS	Arbitrary Inter Frame Spacing
AP	Access Point
APC	Adaptive Power Control
ARIB	Association of Radio Industries and Businesses
ARS	Adaptive (or Automatic) Rate Selection
ASAP	Aggregate Server Access Protocol
ASK	Amplitude shift keying
AWMA	Alternating Wireless Medium Access

B

BER	Bit Error Rate
BPSK	Binary Phase Shift Keying
BRAN	Broadband Radio Access Networks
BRI	Basic Rate Interface
BS	Base Station
BSS	Basic Service Set
BSSID	Basic Service Set Identifier

C

CBC-MAC	Cyclic Block Chaining - Message Authentication Code
CCK	Complementary Code Keying
CCMP	Counter Mode CBC-MAC Protocol
CHAP	Challenge-Handshake Authentication Protocol
CID	Connection or Channel Identifier
CINR	Carrier to Interference Noise Ratio
CRC	Cyclic Redundancy Check
CSCC	Common Spectrum Coordination Channel
CSI	Channel State Information
CSMA/CA	Carrier Sensing Media Access/Collision Avoidance
CSMA/CD	Carrier Sensing Media Access/Collision Detection
CTS	Clear to Send

D

dBi	Decibels relative to an isotropic antenna
dBm	Decibels relative to a 1 mW power level
DBPSK	Differential Binary Phase Shift Keying
DCF	Distributed Coordination Function
DCM	Dual Carrier Modulation

DES	Data Encryption Standard
DFS	Dynamic Frequency Selection
DHCP	Dynamic Host Configuration Protocol
DIFS	DCF Inter Frame Spacing
DLC	Data Link Control
DPSK	Differential Phase Shift Key
DQPSK	Differential Quadrature Phase Shift Keying
DRCA	Distributed Reservation Channel Access
DRP	Distributed Reservation Protocol
DRS	Dynamic Rate Shifting
DS	Distribution System
DSL	Digital Subscriber Line
DSR	Data Set Ready
DSSS	Direct Sequence Spread Spectrum

E

EAP	Extensible Authentication Protocol
EAPoL	Extensible Authentication Protocol over LAN
ECB	Electronic Code Book
ECMA	European Computer Manufacturers Association
EDCA	Enhanced Distributed Channel Access
EDCF	Enhanced Distributed Coordination Function
EDR	Enhanced Data Rate (Bluetooth radio)
EIRP	Equivalent Isotropic Radiated Power
ESS	Extended Service Set
ETSI	European Telecommunications Standards Institute
EUI	Extended Unique Identifier
EWC	Enhanced Wireless Consortium

F

FCS	Frame Check Sequence
FDD	Frequency Division Duplex
FEC	Forward Error Correction
FFI	Fixed Frequency Interleaving
FFT	Fast Fourier Transform
FHSS	Frequency Hopping Spread Spectrum
FIR	Fast Infrared (IrDA)
FSK	Frequency Shift Keying (also Fixed Symmetric Key)
FWA	Fixed Wireless Access

G

GFSK	Gaussian Frequency Shift Keying
GPC	Grant per Connection
GPSS	Grant per Subscriber Station

H

HCCA	HCF Controlled Channel Access
HCF	Hybrid Coordination Function
HCI	Host Controller Interface
H-FDD	Half duplex Frequency Division Duplexing
HIPERLAN	High Performance Radio Local Area Networks
HIPERMAN	High Performance Radio Metropolitan Area Networks
HL/2	HIPERLAN 2

I

I2C	Inter-Integrated Circuit bus
ICMP	Internet Control Message Protocol
ICV	Integrity Check Value
IE	Information Element

IEEE	Institute of Electrical and Electronic Engineers
IETF	Internet Engineering Task Force
IFS	Inter Frame Spacing
IP	Internet Protocol
IPSec	Internet Protocol Security
IR	Impulse Radio
Ir	Infrared
IRAP	International Roaming Access Protocol
IrCOMM	Infrared COM port emulation
IrDA	Infrared Data Association
IrDA Lite	Reduced version of IrDA code
IrLAN	Infrared Local Area Network Protocol
IrLAP	Infrared Link Access Protocol
IrLMP	Infrared Link Management Protocol
IrOBEX	Infrared Object Exchange Protocol
IrTran-P	Infrared Image Exchange Protocol
IrXfer	Infrared File Transfer Protocol
ISDN	Integrated Services Digital Network
ISI	Inter Symbol Interference
IS-IS	Intermediate System to Intermediate System
ISO	International Standards Organisation
ITU	International Telecommunications Union
IV	Initialisation Vector

L

L2CAP	Logical Link Control and Adaptation Protocol
L2TP	Layer 2 Tunnelling Protocol
LAN	Local Area Network

LDPC	Low Density Parity Check
LLC	Logical Link Control
LMDS	Local Multipoint Distribution Service
LMP	Link Manager Protocol
LOS	Line of Sight
LQI	Link Quality Indicator
LSB	Least Significant Bit
LWAPP	Lightweight Access Point Protocol

M

MAC	Media Access Control (also Message Authentication Code)
MAC SAP	MAC Service Access Point
MAN	Metropolitan Area Network
MANET	Mobile Ad-hoc Network
MAS	Media Access Slot
MB-OFDM	Multiband OFDM
MBOA	Multi-band OFDM Alliance
MCF	Mesh Coordination Function
MIC	Message Integrity Check
MIH	Media Independent Handover
MIMO	Multiple Input, Multiple Output
MMC	Micro-scheduled Management Command
MPDU	MAC Protocol Data Unit
MSB	Most Significant Bit
MS-CTA	Micro-scheduled Channel Time Allocation
MSDU	MAC Service Data Unit
MTU	Maximum Transmission Unit

N

NAT	Network Address Translation
NFC	Near Field Communications
NLOS	Non Line-of-Sight
NOS	Network Operating System
NRZI	Non Return to Zero Inverted

O

OCB	Offset Code Book
OFDM	Orthogonal Frequency Division Multiplexing
OFDMA	Orthogonal Frequency Division Multiple Access
OSI	Open System Interconnect
OSPF	Open Shortest Path First

P

PAN	Personal Area Network
PAM	Pulse Amplitude Modulation
PAP	Password Authentication Protocol
PAT	Port Address Translation
PBCC	Packet Binary Convolution Coding
PCA	Prioritised Contention Access
PCF	Point Coordination Function
PCMCIA	Personal Computer Memory Card International Association
PDU	Protocol Data Unit
PER	Packet Error Rate
PHY	Physical Layer
PIFS	PCF Interframe Spacing
PKI	Public Key Infrastructure
PMP	Point to Multipoint

POA   Point of Attachment

POS   Personal Operating Space

PPP   Point to Point Protocol

PRI   Primary Rate Interface

PRN   Pseudo Random Noise

PSM   Pulse Shape Modulation

PTK   Pair-wise Temporal (or Transient) Key

Q

QAM   Quadrature Amplitude Modulation

QoS   Quality of Service

QPSK   Quadrature Phase Shift Keying

R

RADIUS   Remote Authentication Dial-In User Service

RC4   Rivest (or Ron's) Code 4

RF   Radio Frequency

RFC   Request for Comments

RFID   Radio Frequency Identification

RIP   Routing Information Protocol

RLC   Radio Link Control

RSA   Rivest Shamir Adleman

RSN   Robust Security Network

RSSI   Received Signal Strength Indicator

RTS   Request to Send

RZI   Return to Zero Inverted

S

SCO   Synchronous Connection-Oriented

SDM   Space Division Multiplexing

SDMA	Space Division Multiple Access
SDU	Service Data Unit
SEEM	Simple, Efficient and Extensible Mesh
SIFS	Short Interframe Spacing
SIR	Serial IrDA
SMB	Server Message Block
SNMP	Simple Network Management Protocol
SNR	Signal to Noise Ratio
SoHo	Small Office Home Office
SOP	Simultaneously Operating Piconets
SPI	Serial Peripheral Interface
SPIT	Spam over Internet Telephony
SS	Subscriber Station
SSID	Service Set Identifier
ST(B)C	Space Time (Block) Coding

T

TC	Traffic Class
TCP/IP	Transport Control Protocol / Internet Protocol
TDD	Time Division Duplex
TDM	Time-Division Multiplexed
TDMA	Time Division Multiple Access
TFC	Time Frequency Code
TFI	Time Frequency Interleaving
TG	Task Group
Tiny-P	Flow control protocol for IrLMP connections
TKIP	Temporal Key Integrity Protocol
TLS	Transport Layer Security

TTLS	Tunnelled TLS
TPC	Transmitter Power Control
TSN	Transition Security Network
TXOP	Transmit Opportunity

U

UART	Universal Asynchronous Receiver Transmitter
UDP	User Datagram Protocol
U-NII	Unlicensed - National Information Infrastructure
USB	Universal Serial Bus
UTP	Unshielded Twisted Pair
UWB	Ultra Wide Band

V

VFIR	Very Fast Infrared (IrDA)
VLAN	Virtual Local Area Network
VoIP	Voice over Internet Protocol
VoWLAN	Voice over Wireless Local Area Network
VPN	Virtual Private Network

W

WAE	Wireless Application Environment
WAP	Wireless Application Protocol
WDS	Wireless Distribution System
WEP	Wired Equivalent Privacy
WIGWAM	Wireless Gigabit With Advanced Multimedia Support
WiMAX	Worldwide Interoperability for Microwave Access
WISP	Wireless Internet Service Provider
WLAN	Wireless Local Area Network

WMAN	Wireless Metropolitan Area Network
WMM	Wi-Fi Multimedia
WPA	Wi-Fi Protected Access
WPAN	Wireless Personal Area Network
WRAN	Wireless Regional Area Network
WRAP	Wireless Robust Authenticated Protocol
WUSB	Wireless USB

X

xDSL	generic Digital Subscriber Line (ADSL, etc.)

Networking and Wireless Networking Glossary

A

Access Point: A wireless networking device that acts as a wireless hub and typically connects a wireless LAN to a wired network or to the Internet.

Ad-hoc Mode: A wireless networking mode, also referred to as peer-to-peer mode or peer-to-peer networking, in which wireless enabled devices communicate with each other directly, without using an access point as a communication hub. See also Infrastructure Mode.

Adaptive Burst Profiling: In adaptive burst profiling, transmission parameters such as modulation and coding schemes are adjusted on a frame by frame basis to meet the needs of individual client stations. This is a key element of the 802.16 MAC that provides flexibility for a wide range of services and devices in metropolitan area networking.

Adaptive Frequency Hopping (AFH): Adaptive Frequency Hopping limits the channels used by a frequency hopping spread spectrum (FHSS) device to avoid channels being used by other co-located devices. AFH is applied in the 2.4 GHz ISM band to enable Bluetooth and Wi-Fi devices to co-operate without interference.

Asymmetric Digital Subscriber Line (ADSL): ADSL is one of the number of technologies that enables higher bandwidth over standard copper telephone lines. ADSL can achieve data rates of 640 kbps in the uplink and 9 Mbps in the downlink direction, with a range of 6 km from the phone exchange.

Asynchronous: In an asynchronous service, such as a generic file transfer, the transmission of packets within a data stream can be separated by random intervals. This compares with the more stringent time requirements of isochronous or synchronous services (q.v.).

B

Backhaul: Backhaul refers to the transmission of network traffic from a remote site, such as a wireless MAN base station, back to an ISP or other Internet point-of-presence. A DSL link might provide backhaul for an urban Wi-Fi hotspot, while a remote rural network might require a long-range wireless or satellite link.

Bandwidth: Bandwidth is a measure either of the frequency width of a signal (measured in Hertz), or the total amount of data that can be transmitted in a certain period of time over a particular medium or using a particular device (measured in bps). The bandwidth of a transmitted signal is measured between the frequencies at which the signal drops to half its peak power (the 3dB points).

Barker Code: A Barker code is a binary sequence that has low correlation with a time-shifted version of itself (low auto-correlation). The 11-bit Barker code (10110111000) is used as the spreading or chipping code in 802.11 DSSS.

Baseband: Baseband refers to the signal in a communication system before it is modulated and multiplexed onto its carrier. The term is commonly used to refer both to the signal and to the hardware/software that processes the signal.

Beamwidth: The coverage angle of a radio antenna, ranging from 360 degrees for an omnidirectional antenna to a narrow pencil beam with a high gain directional antenna such as a Yagi or parabolic dish.

Binary Phase Shift Keying (BPSK): A modulation technique, using two phases of the carrier to represent data symbols 1 and 0.

Bluetooth: A wireless PAN technology used for voice and data links with a typical range of 10 metres. Bluetooth delivers a standard data rate of 720 kbps using frequency hopping spread spectrum in the 2.4 GHz ISM radio band. With extended data rate (EDR) 2 Mbps and 3 Mbps rates can be achieved.

Bonding: The process of link creation, pairing and authentication that occurs between Bluetooth devices.

Bridge: A bridge is a link between two networks, for example a point-to-point wireless link connecting two wired networks.

Broadcast: A broadcast message is transmitted to all receivers or stations connected to the network, in contrast to multicast or unicast messages (q.v.). Beacon messages used in many types of wireless networks are an example of broadcast messages.

C

Carrier Sensing Media Access/Collision Avoidance: CSMA/CA is a method for multiple users to share access to the wireless medium while avoiding interfering with each other. A transmitter waiting to send data senses the medium to see if another station is transmitting, and uses a variety of strategies, such as random back-off or RTS/CTS messages, to avoid collisions with other transmitters when the medium becomes free.

Chipping Code: A code, such as a Barker or complementary code, used to spread a single bit into a longer sequence of chips to enable detection in a noisy communication channel.

Coding Rate: The coding rate is an indication of the error-correction code overhead that is added to a data block to enable error recovery on reception. It is equal to $m/(m+n)$ where n is the number of error correction bits applied to a data block of length m bits. Efficient error correction codes keep the coding rate close to unity.

Complementary Code Keying (CCK): Complementary Code Keying is a type of Direct Sequence Spread Spectrum in which complementary codes (typically a set of 64 specific bit patterns) are used to encode the data stream and provide processing gain to enable detection of weak signals in a noisy environment.

Connection Oriented/Connectionless Communication: Connection oriented communication takes place over a connection established between the Logical Link Control layers in the sending and receiving devices.

Link flow and error control mechanism are available to ensure reliable and error free delivery for connection oriented communication. In contrast, connectionless communication proceeds without an LLC to LLC connection, when flow or error control are not required.

Cyclic Redundancy Check: A cyclic redundancy check is the computation for a block of data of a number, commonly called the checksum, which summarises and represents the content and organisation of bits in the input data block. By re-computing the CRC, the receiving station can detect any change, whether due to random transmission errors or to malicious interception. In a CRC, the input data bits are taken as the coefficients of a polynomial. This polynomial is divided by another fixed polynomial and the CRC checksum bits are the coefficients of the remainder polynomial. The fixed divisor polynomial used in CRC-32 is $x^{32} + x^{26} + x^{23} + x^{22} + x^{16} + x^{12} + x^{11} + x^{10} + x^8 + x^7 + x^5 + x^4 + x^2 + x + 1$, which can be represented in hexadecimal as either 04C11DB7 or EDB88320 depending on the bit-order convention (LSB first or MSB first).

D

dBi: A logarithmic measure of the gain of an antenna relative to an isotropic antenna.

dBm: A logarithmic measure of power relative to 1 milliwatt (mW). A power level of P dBm is ten to the power ($P/10$) milliwatts, so that 20 dBm is 100 mW.

Delay Spread: The delay spread is the variation in arrival time of wireless signals that propagate from a transmitter to a receiver over multiple paths. Inter symbol interference (ISI) will occur if consecutive symbols are transmitted closer together in time than the delay spread.

Differential Binary Phase Shift Keying (DBPSK): A variation of the BPSK modulation technique in which a symbol is represented by a change of carrier phase rather than an absolute phase value.

Differential Quadrature Phase Shift Keying (DQPSK): A variation of QPSK modulation in which a symbol is represented by a change of phase between two of the 4 QPSK constellation points, rather than the absolute phase value of one point.

Digital Subscriber Line: A class of high speed Internet connections using standard telephone lines and delivering data rates of up to 1.5 Mbps.

Directional Antenna: A directional antenna focuses transmitted power into a narrow beam, increasing the range at the expense of reduced angular coverage. Patch, Yagi and parabolic are types of directional antennas.

Direct Sequence Spread Spectrum (DSSS): Direct Sequence Spread Spectrum is a data encoding technique in which the input bit stream is XOR'd with a chipping code to increase its bandwidth. On reception, the chipping code sequence is easier to detect in a noisy environment than a single transmitted bit, resulting in an additional gain in the system known as the processing gain. Longer chipping codes result in higher processing gain.

Diversity: Diversity refers to a technique for improving the transmission of a signal, by receiving and processing multiple versions of the same transmitted signal. The multiple received versions can be the result of signals following different propagation paths (spatial diversity), being transmitted at different times (time diversity) or frequencies (frequency diversity).

The simplest example of diversity in practice is the use of diversity antennas, where a receiver continuously senses the strength of received signals on two or more antennas and automatically selects the antenna receiving at maximum strength.

Dynamic Host Configuration Protocol (DHCP): DHCP is a protocol that automatically provides network addressing and configuration information such as an IP address, sub-net mask and default gateway to a device when it connects to the network. Typically a device retains an assigned IP address for a specific administrator-defined period of time known as the lease period.

Dynamic Frequency Selection (DFS): Dynamic Frequency Selection enables wireless networking devices to select a transmission channel to be used in order to avoid interference with other users, particularly radar and medical systems. DFS was introduced in the 802.11h supplement to enable 802.11a WLANs to comply with European regulations.

Dynamic Routing: In contrast to static routing (q.v.), dynamic routing is the process whereby routers continuously update and exchange routing information in order to automatically adjust to changes in network topology.

E

Equivalent Isotropic Radiated Power (EIRP): EIRP measures the total effective transmitted power of a radio, including antenna gain and any cable and connector losses between transmitter and antenna. A 100 mW (20 dBm) transmitter with a 4 dB antenna and 1 dB of losses will have an EIRP of 200 mW (23 dBm).

Ethernet: Ethernet is the predominant wired networking technology, developed at the Xerox Palo Alto Research Centre in the 1970s and standardised in the IEEE 802.3 specification. Ethernet types are designated a ABase-B, where A specifies the data rate, now up to 10 Gbps, and B specifies the cabling type, examples being T for twisted pair copper cabling and SX for LED powered multi-mode optical fibre.

Extended Unique Identifier: An IEEE alternative to MAC addresses, extended in length from 48-bit to 64-bit addresses in EUI-64.

F

Fading: Fading or multipath interference occurs when a primary signal combines with delayed signals, typically caused by reflection or refraction from objects on or near the line-of-sight, resulting in constructive or destructive (fading) interference or phase shifts. Fading can be identified and corrected by techniques such as the use of pilot tones in OFDM radios.

Firewall: A firewall is a software component that controls the external interfaces of a network and restricts or blocks certain types of traffic or activity. Firewalls are essential for network security, and careful configuration is essential to ensure correct operation and to avoid unintended interruption of authorised network traffic.

Forward Error Correction: Forward error correction (FEC) is a method of reducing bit errors during data transmissions. Redundant bits, each usually a complex function of many bits from the input data stream, are added to the transmitted data and enable the receiving device to detect and correct a certain fraction of bit errors that occur during transmission. The fraction of original data bits in the final transmitted stream is called the coding rate.

Frequency Reuse Factor: The frequency reuse factor is a measure of the degree to which a certain frequency can be used throughout the network, and therefore is a measure of the extent to which the full RF transmission bandwidth is available to carry data. For a large-scale 802.11b/g WLAN, with 3 non-overlapping channels patterned across adjacent access points, the frequency reuse factor is 1/3, as each channel is only used in 1/3 of the operating area of the network.

Frequency Hopping Spread Spectrum (FHSS): Frequency Hopping Spread Spectrum is a spread spectrum technique, used in Bluetooth devices, in which the centre frequency of the modulated carrier hops periodically between a number of predetermined frequencies. The sequence of frequencies used is determined by a hopping code which is also known to the receiver.

Fresnel Effect: A phenomenon in radio wave propagation in which an object that does not directly obstruct the line-of-sight from transmitter to receiver still causes attenuation of a transmission. The impact depends on how close the object is to the line-of-sight.

G

Gateway: A gateway is a network component that connects one network to another, typically to the Internet. The gateway performs routing and protocol translation (e.g. NAT, VPN passthrough), and may also provide additional functions such as acting as a DHCP server.

H

Hub: A hub is a network device that provides a central connection point for other devices. In contrast to a switch (q.v.), a hub broadcasts each data packet to every connected device, sharing the available bandwidth among all devices in the network.

I

Infrastructure Mode: Infrastructure Mode is a mode of operation of a wireless network in which communication between devices takes place via an access point rather than directly between devices, as occurs in peer-to-peer or ad-hoc mode (q.v.).

Initialisation Vector (IV): An initialisation vector is part of the key used in an encryption algorithm, and is typically changed for each data packet. The IV is added to a secret key, such as a 40-bit key derived from a WEP passphrase, to obtain the full key used in the encryption algorithm. The IV is transmitted with the encrypted message and the receiving station, knowing the secret key, can determine the full encryption key. The IV prevents the occurrence of patterns in the encrypted data that would otherwise make it easier for a hacker to determine the secret key.

Internet Protocol (IP): Internet Protocol is the Network layer protocol that provides addressing and routing functions on the Internet. The current version is IPv4.

IP Address: A number that uniquely identifies a device on the Internet so that other devices can communicate with it. IPv4 uses 32-bit addresses, while IPv6 addresses will be 128-bit. IP addresses are used by protocols at Layer 3 and above, and are translated into MAC addresses by the address resolution protocol (ARP) for use at Layer 2 and below.

ISM (Industrial, Scientific and Medical): The ISM bands are three radio bands at 900 MHz, 2.4 and 5.8 GHz that were originally reserved for licence exempt use in industrial, scientific and medical applications, but are now also home to the main wireless networking PHY layer technologies.

Isochronous: An isochronous service requires data to be delivered within certain time constraints. Multimedia streams require an isochronous transport service to ensure that data frames are delivered as fast as they are displayed. The requirements of an isochronous service are not as rigid as those of a synchronous service, but are also not as lenient as an asynchronous service (q.v.).

J

Jitter: Jitter refers to the variability in latency of individual transmitted packets, and is particularly important in determining quality of service for isochronous data services such as streaming video.

L

Latency: The time taken by a data packet to travel from its source to its destination. Latency is particularly important in some network services such as voice and video streaming, where transmission delays can seriously reduce the quality of service to the end user.

Line-of-Sight (LOS): Line-of-sight refers to an unobstructed line between transmitting and receiving stations. Line-of-sight is required for any wireless link for frequencies above ca. 11 GHz. When setting up long distance wireless links, a line-of-sight survey can be used to assess whether one aerial can be "seen" by another.

Link Margin: The excess available power, over and above that required by the link budget, to achieve a desired received signal strength and SNR at the receiver.

Local Area Network (LAN): A network used to link computers and other devices such as printers over short distances, typically tens of metres to a few hundred metres. The most common LAN technologies are 100 BaseT Ethernet for wired and 802.11b/g or Wi-Fi for wireless networks.

Low Density Parity Check Codes: LDPC codes are error correction codes that are computationally efficient and perform close to the theoretical limit in enabling error recovery in noisy communication channels. A check code is computed from a data block by sparsely sampling bits in the block using a random sampling matrix. For example, the first bit in the check code may be defined as $y_1 = x_1 + x_3 + x_8 + x_9 + x_{10}$, where the sum is modulus 2. The check code enables lost bits in the received data block to be recovered with high efficiency.

M

MAC Service Access Point (MAC SAP): The MAC SAP is the logical point at the "top" of the MAC layer, where the MAC interfaces with the Network layer and higher layers in the OSI model. The MAC SAP is taken as a reference point when specifying the effective data rate of a network, as distinct from the data rate transmitted at the PHY layer.

Manchester Coding: Manchester code is a data encoding method in which each bit or chip of data is represented by a transition between two states, commonly at the middle of the bit or chip period. The transition may be between amplitude, frequency or phase states.

Media Access Control (MAC) Address: The MAC address is a unique hardware address of a network interface device and is used by Layer 2 protocols to address data packets to a desired destination device. MAC addresses are 48-bits long, although a 64-bit Extended Unique Identifier has been developed by the IEEE.

MICHAEL (MIC): The message integrity check (MIC) is a feature of the wireless protected access (WPA) enhancement to IEEE 802.11 security, developed by the Wi-Fi Alliance as an interim improvement on WEP in advance of the IEEE 802.11i standard.

Miller Coding: Miller coding is a type of Manchester coding (q.v.) where a 1-bit is represented by a transition of some type (amplitude, frequency, phase) at the centre of the bit period and a 0-bit is represented by no transition. A repeated 0-bit is represented by a transition at the start of the bit period. In modified Miller coding, each transition is replaced by a negative pulse.

Modulation: Modulation is the technique used to combine a digital data stream with a transmitter's carrier signal.

Modulation Index: The modulation index is defined as $[A-B] / [A+B]$ where A and B are the maximum and minimum signal amplitudes.

Multicast: A multicast message is transmitted to a subset of all stations on the network, in contrast to broadcast and unicast messages (q.v.).

Multipath Interference: See Fading

Multiple Input Multiple Output (MIMO): Multiple Input Multiple Output refers to wireless links with multiple transmitter and receiver antennas. Processing signals from multiple antennas can increase the available data bandwidth or reduce the average bit error rate (BER) of the link, by exploiting spatial diversity (spatial multiplexing).

N

Network Address Translation (NAT): Network Address Translation is the replacement of a local network address with an address that can be used on an external network such as the Internet, and is performed by the gateway device that connects the two networks.

Nonce: A random number used once and then discarded, for example as the seed for an algorithm used to determine encryption or authentication keys.

O

Omnidirectional Antenna (Omni): An omnidirectional antenna has a 360 degree beamwidth in the horizontal plane (perpendicular to the axis of a vertically mounted antenna), and is used when transmission and reception is needed in all radial directions. The gain of an omni can be increased at the expense of reducing the vertical beamwidth.

Orthogonal Frequency Division Multiplexing (OFDM): OFDM is a multiplexing technique in which a data stream is split into a number of parallel streams of lower bit rate which are modulated onto a set of subcarriers and transmitted concurrently. The characteristic of orthogonality refers to the frequency separation of the subcarriers which is chosen to reduce inter-carrier interference.

P

Packet Binary Convolution Coding (PBCC): Packet Binary Convolution Coding is a data coding and modulation scheme developed by Texas Instruments as a proposal for the 802.11g standard. OFDM was selected in preference to PBCC as it delivers 54 Mbps compared to the 33 Mbps achievable with PBCC.

Pairing: The initial step in creating a link between Bluetooth devices when a PIN is entered into both devices.

Pair-wise Temporal (or Transient) Keys: Encryption keys that are derived as part of a device authentication process and used by both devices to encrypt data traffic for the duration of the connection.

Personal Computer Memory Card International Association (PCMCIA): The Personal Computer Memory Card International Association was formed in 1989 to define standards for portable computer expansion cards. PCMCIA cards are now known as PC cards.

Piconet: An ad-hoc collection of devices in a PAN with one device acting as a master and the others as slaves. The master device sets the clock and other link parameters such as a FHSS hopping pattern. In Bluetooth piconets, each master can connect to 7 active or up to 255 inactive (parked) slave devices.

Point of Attachment (POA): A point of attachment is a generic point at which wireless device connects to a wireless network. Wi-Fi access points and WiMAX or 3GPP cellular base stations are all examples of points of attachment.

Point of presence (POP): A point of presence (POP) is an access point into the Internet. An ISP or other online service provider will have one or more POPs to carry customer traffic onto the Internet.

Point to Point Transport Protocol (PPTP): A protocol that allows secure transmission of data over virtual private networks (VPN) that make use of insecure connections such as the public telephone system or the Internet.

Processing Gain: Processing gain is achieved when a chipping code is used to spread a bit stream into a wider bandwidth chip stream. The processing gain in decibels is equal to $10 \log_{10}(C)$ where C is the length of the chipping code.

Protocol: A standard set of rules, or language, for enabling network communication between devices. Examples of protocols include HTTP, FTP, TCP and IP.

Q

Quadrature Phase Shift Keying (QPSK): A modulation technique in which an input data symbol is represented by one of four phase states of the carrier wave.

R

Receiver Sensitivity: The receiver sensitivity is a measure in dBm of the weakest signal that a receiver can reliably decode at a specified bit error rate (BER). Receiver sensitivity is a function of the modulation method used, since more complex schemes, such as 16- or 64-QAM, require a higher received signal strength for reliable decoding. Typically, decoding 64-QAM requires 15–20 dB higher signal strength than decoding BPSK.

Request for Comments (RFC): A publication by an industry body requesting comments on a proposal relating to standards or generic solutions to a technical problem. The Network Working Group publishes some of the seminal RFCs that established the fundamentals of much of the networking and Internet technology described in this book. Some important examples are;

RFC 791   Internet Protocol

RFC 1738   Uniform Resource Locators (URL)

RFC 1945   Hypertext Transfer Protocol (HTTP 1.0)

The RFC archive at www.faqs.org/rfcs is a fascinating history of the development of these key technologies.

Router: A router is a network device that examines the IP address within a data packet and forwards the packet on towards its destination. The sharing of information between routers about paths to other network destinations is achieved using protocols such as Router Information Protocol (RIP), Open Shortest Path First (OSPF) and Intermediate System to Intermediate System (IS–IS).

RSA public key algorithm: The RSA algorithm is named after cryptographers Rivest, Shamir and Adleman, and is used in public key cryptography to establish a public plus private key pair. The algorithm starts with two large primes, p and q, whose product $n = pq$. A number e is chosen that is less than n and such that e and $(p-1)(q-1)$ have no common factor except 1. Another number d is found such that $(ed - 1)$ is divisible by $(p-1)(q-1)$. The public key is the pair (n, e), the private key is (n, d).

The security of the RSA algorithm is based on the difficulty of factoring a product of two large prime numbers. If this could be done then the private key could be computed from the public key.

S

Scatternet: A scatternet is the linking of multiple co-located Bluetooth piconets through the sharing of a common device. A device in a scatternet can be both a master in one piconet and a slave in another.

Service Set Identifier (SSID): The SSID identifies a group of wireless devices networked through a single access point. Data packets transmitted between these devices are identified by carrying the SSID, and will be ignored by devices operating under a different SSID.

Simple Network Management Protocol (SNMP): Simple Network Management Protocol is a communications protocol that enables the configuration and monitoring of network devices such as access points and gateways. SNMP is used to perform network management functions such as configuring security and access policies (e.g. MAC filtering) and for network traffic monitoring.

Spectral Efficiency: A measure, in bits/Hz, of the efficiency of bandwidth used for transmitting information. Increasing modulation complexity (for example, from BPSK to 64-QAM) increases spectral efficiency at the cost of higher received signal strength required for successful demodulation.

Spread Spectrum: A radio frequency data transmission technique that spreads the signal over a wide band of frequencies. The processing gain on decoding results in the signal being less susceptible to noise and to interference from narrow band radios operating at similar frequencies.

Static Routing: In contrast to dynamic routing (q.v.), in static routing a fixed address table is used to route network traffic and requires manual intervention for updates in response to changes in network topology.

Switch: A network device, similar to a hub, which actively switches data packets to the sub-net on which the destination device is located.

Synchronous: In a synchronous service, data bursts within a continuous data stream must be delivered at specific time intervals.

T

Temporal Key Integrity Protocol (TKIP): Part of the WPA security enhancement defined by IEEE 802.11i. TKIP manages the sharing and updating of encryption keys between access points and associated wireless stations.

Transport Control Protocol (TCP): TCP manages the transport of messages over a network, including the fragmentation into packets, in sequence reconstruction from received packets, error checking and requests to retransmit missing packets.

Tunnelling: Tunnelling creates a secure link across an unsecured network by encapsulating data packets structured according to one protocol with a wrapper defined by another protocol. Tunnelling involves three different protocols; a carrier protocol used by the network that the data is transmitted over (e.g. IP), an encapsulating protocol that provides the wrapper and the passenger protocol that structures the original data packet.

Turbo codes: Turbo codes are error correction codes that rely on the computation of two independent parity checks on each data block. On decoding, the two check codes are used to independently verify the received data block. If the two decoders disagree on the data block an iterative exchange of information takes place until they converge on a single solution.

U

User Datagram Protocol (UDP): UDP is a data transport protocol that performs similar functions to TCP, but does not check for the arrival of all data packets or request retransmission if any packets are missed. UDP is used in applications such as VoIP or media streaming, where recovery of missed packets is of no value and they are simply dropped.

Unicast: A unicast message is transmitted to a single identified receiving station, in contrast to multicast and broadcast messages (q.v.).

Unshielded Twisted Pair (UTP): Unshielded twisted pair is the most common type of cabling for wired Ethernet connections and consists of four twisted copper wire pairs, terminated with an RJ45 connector. Different categories of UTP cabling are defined for different network speeds from Cat 1 for low speed (1 Mbps), Cat 5 the most common Ethernet cabling for 100 Mbps, up to Cat 7 for ultra-fast Ethernet at 10 Gbps.

Ultra Wide Band (UWB): Ultra Wide Band is defined as any transmission in which the 3 dB bandwidth is 20% or more of the centre frequency, with a minimum bandwidth of 500 MHz. In the USA, the FCC defines a UWB passband from 3.1 to 10.6 GHz in which maximum transmitter power must be below -41.3 dBm/MHz. This power level is less than that permitted for unintentional emitters such as computers and other electronic devices.

V

Voice over Internet Protocol (VoIP): Voice over Internet Protocol refers to the transmission of voice data over the Internet using the UDP transport protocol.

Virtual LAN (VLAN): A virtual LAN is a subset of client stations in a LAN or WLAN that is defined using software in order to be able to isolate traffic in the VLAN from the wider LAN. For example this may be because traffic in the VLAN is more susceptible to attack (e.g. VoIP phones) and needs to be treated as "untrusted" to assure the security of the rest of the LAN.

Virtual Private Network (VPN): A Virtual Private Network is formed when a remote device connects to a private network via a public network such as the Internet. A tunnelling protocol such as IPSec, L2TP or PPTP ensures privacy by encrypting data transferred over the public network.

W

Wired Equivalent Privacy (WEP): A security feature of the 802.11 standard that was intended to provide security equivalent to that of a wired LAN.

Encryption vulnerabilities led to WEP being superseded by WPA and by 802.11i security mechanisms.

Wireless Ethernet Compatibility Alliance (WECA): Original industry sponsors of the Wi-Fi standard, now re-branded as the Wi-Fi Alliance. The Wi-Fi logo gives a guarantee of interoperability of 802.11 based wireless networking devices.

Wi-Fi Protected Access (WPA): An enhanced version of IEEE 802.11 security, developed by the Wi-Fi Alliance as a precursor to the IEEE 802.11i standard, which provides stronger encryption and manages the distribution of encryption keys. WPA also provides user authentication and message integrity checking (MICHAEL).

Wireless Personal Area Network (WPAN): A network for communication and inter-connection of devices within a personal operating space, with emphasis on short-range, low power and low cost.

Wireless Internet Service Provider (WISP): A public Internet service provider using wireless network connections to subscribers.

Wireless Local Area Network (WLAN): A wireless network that uses licensed or unlicensed radio frequencies such as the 2.4 or 5 GHz ISM bands to connect wireless enabled computers or other devices in a local area extending typically over 10–100 metres.

X

X-10: A power line based home automation technology, primarily used for lighting control.

Z

ZigBee: A wireless networking technology based on the IEEE 802.15.4 PHY and MAC standard, and aimed at short range, very low power and low data rate monitoring and control applications.

Index